THE BREEDER REACTOR

Proceedings of a Meeting at the
University of Strathclyde
25 March 1977

SIR SAMUEL CURRAN
LORD HINTON OF BANKSIDE
SIR JOHN HILL
C. W. BLUMFIELD
HUGH C. SIMPSON
K. J. W. ALEXANDER
D. W. HUNT
D. W. CLELLAND
J. H. FREMLIN
D. R. BERRIDGE & K. R. VERNON
F. L. TOMBS

Edited by
J. S. FORREST F.R.S.

Sponsored by
The University of Strathclyde
and
The Highlands and Islands Development Board

1977
SCOTTISH ACADEMIC PRESS
EDINBURGH

Published by
Scottish Academic Press Ltd,
33 Montgomery Street
Edinburgh EH7 5JX

SBN 7073 0216 1

COVER

Dounreay Fast Reactor, 1959–1977
from a photograph by courtesy of
United Kingdom Atomic Energy Authority

Printed in Great Britain by
R. & R. Clark Ltd, Edinburgh

Contents

Abbreviations

AGR	Advanced Gas-cooled Reactor.
DFR	Dounreay Fast Reactor.
PFR	Prototype Fast Reactor.
CFR	Commercial Fast Reactor.
FBR	Fast Breeder Reactor.
LMFBR	Liquid Metal Fast Breeder Reactor.
HTR	High Temperature Reactor.
BWR	Boiling Water Reactor.
PWR	Pressurised Water Reactor.
SGHWR	Steam Generating Heavy Water Reactor.
ICRP	International Commission on Radiological Protection.
NII	Nuclear Installations Inspectorate.
mtce	Million tonnes coal equivalent.
Curie	quantity of radioactive isotope undergoing $3{\cdot}7 \times 10^{10}$ disintegrations per second.
Roentgen	amount of radiation producing one electrostatic unit of ions per cubic centimetre.
rad	unit of radiation dose. 1 rad is equal to an energy absorption of 100 ergs/g.
rem	roentgen equivalent man – the unit of body exposure to radiation.

Foreword

This meeting had its origin in a conversation between Sir Samuel Curran and myself in which we were expressing some dissatisfaction and disquiet at the emotive way in which advanced technology is being presented to the public and important decisions postponed pending 'public debate'. The greatness of Britain was founded on engineering and technology; for example, on the development of our coal and steel industries – and we didn't ban steel-making because it could lead, as it undoubtedly did lead, to the 'proliferation' of weapons all over the world.

The maintenance of our standard of living depends vitally on the fostering of our innovative and inventive genius right through to the stage of manufacturing technical equipment and systems which can be sold competitively throughout the world. A good example of our technological achievement is the British nuclear power programme, based on British reactors, and also the successful development of the breeder reactor. It will be foolhardy if we do not start building a commercial breeder reactor station now to pave the way to safeguard our future energy supplies.

It was with such thoughts in mind that we felt that it would be useful to provide a forum for a meeting on the breeder reactor at which the subject would be presented by speakers who have had first-hand experience of nuclear energy for many years and had held, or were holding, responsible positions in organisations carrying-out the nuclear energy programmes.

A second reason for the meeting was to provide a Scottish venue, which is surely appropriate because of Scotland's involvement at Dounreay, and the keenness of the Highlands and Islands Development Board to maintain, and in fact increase, this activity from a social as well as a technical point of view. Nearly 30% of the electricity in Scotland is generated in nuclear stations, at less cost than in coal- or oil-fired stations, and Scotland clearly wishes to maintain such a leading position.

When in October 1976, we decided to hold the meeting in March 1977, we were worried that perhaps by that time the meeting might lose its impact, as a decision to build the first commercial breeder reactor might have already been made – we needn't have worried!

JOHN S. FORREST
Meeting Organiser

Setting the Scene

SIR SAMUEL CURRAN, F.R.S.

Principal and Vice-Chancellor, University of Strathclyde

SUMMARY

The reasons for the special meeting on the breeder reactor are outlined with some reference to the special Scottish interest in the topic. Approximately 30% of the electrical energy generated in Scotland is nuclear and the special developments at Dounreay make policy decisions on the future of the commercial breeder reactor urgent. The participants review the major questions arising in arriving at such decisions. In effect an attempt is made to respond to the wish of the Secretary of State for Energy to have informed debate. To set the scene the importance of energy availability as regards the strength of the national economy is stressed and the reasons for an increasing energy demand put forward. Examination of alternative sources of energy shows that none is definitely capable of filling the foreseen energy gap. This implies an integrated thermal/breeder reactor programme as the way to close the anticipated gap. The problems of disposal of radioactive waste and the safeguards in the handling of plutonium are outlined. Longer-term benefits, including the consumption of plutonium and naturally occurring radioactive materials, are examined.

SOME REASONS FOR THE MEETING

We of the University of Strathclyde are happy to act as hosts at this special Meeting on the Breeder Reactor and happy that our co-sponsors are the Highlands and Islands Development Board. A few remarks on the reasons for organising the meeting are pertinent.

First, it was believed important that particularly Scottish interests in nuclear energy and the breeder reactor programme should be explored. With approximately 30% of the electrical energy generated in Scotland nuclear in origin, interest in the future development of fission energy is clearly of great importance. It might be that the 30% is the largest fraction in the world at present. Second, the only fast breeder reactors generating substantial amounts of energy are in the North of Scotland. At Dounreay the Experimental Fast Breeder has just been closed down by Lord Hinton after the completion of a long and successful period of research and development based on its use. The prototype fast breeder (PFR) is in operation at Dounreay; it has been successfully brought into commission and should soon be supplying steadily its 250 MW (electrical) to the grid. These are very notable achievements and together they represent a considerable triumph for scientists and engineers of the United Kingdom Atomic Energy Authority working at Dounreay. The virtually uninterrupted progress of work over the years illustrates once again a level of achievement in research, and also in development, of which we can be proud. We sometimes accuse ourselves in this country of failing to turn our research success into industrial advances. In the case of the breeder reactor however it would appear to be the fact that the scientists and engineers are apparently convinced that the commercial fast reactor (CFR) should be constructed and the first commercial scale generating station (around 1300 MW(e)) shown to be practicable as the first stage towards an integrated thermal and fast breeder system well suited to meet a substantial part of the energy needs of the U.K. within some twenty to thirty years.

I shall mention the fact that arguments about the so-called plutonium economy have been expressed. It is hoped in this Meeting on the Breeder Reactor to help to clarify many of the issues and to establish the facts on which a decision regarding the construction of CFR should be based.

The language of some opponents of an expansion of the fission energy programme has been somewhat emotive. It is hoped that this Meeting will present an unbiased view of the subject and one pertinent to the British scene. A number of reports by experts have appeared and it is sometimes forgotten that some of the most authoritative have little or no relevance to the U.K. situation. Thus, while it may be reasonable to go slow on the breeder development in the U.S.A. which has large uranium resources (both within and without the U.S.A.) and a large thermal reactor construction programme (for example the PWR is exported in numbers), such a go-slow policy may be inappropriate and possibly extremely harmful to the economy of the U.K. Each of the advanced countries has to evaluate the energy situation in the light of its own situation. The same argument applies to questions of fuel reprocessing. Thus it is difficult to imagine that Japan can tolerate almost complete dependence on oil and gas; having to import almost 98% of its energy it is certain that the construction of reactors will be seen to be critically important. If we do not reprocess the reactor fuel the Japanese will be forced to do it elsewhere, possibly in Japan. With our proven ability to reprocess fuel it is difficult to see why it should not be a profitable industry for the U.K. It will be shown that it does not result in the creation of huge amounts of radioactive waste, and phrases like 'nuclear dustbins' do not help in arriving at a rational policy.

Most of us realise that there has already been a good deal of debate about fission energy and its future as part of the total programme for energy production in the U.K. There have been meetings on energy held elsewhere and at many of these, discussion has centred largely on our British nuclear programme. The effect of public debate in which experts took part has been explained by Edward Teller in a recent talk in London. He stressed that in California public opinion changed from anti-nuclear to 67% for nuclear in the referendum (see Atom 244, February 1977, 14). The Secretary of State for Energy, Mr. Wedgwood Benn, has asked that there should be debate on the energy issues and this was one of the reasons for organising this Meeting.

Today we have present with us a number of people with long experience and deep knowledge of the nuclear industry and who will speak to us on important aspects of an important topic. Some are authors of important books and papers. Some have experience and no personal commitment to particular energy programmes: it is hoped that for that reason they can speak impartially on this subject. I trust that I fall myself into this second category, as one who has spent a good many years in nuclear work but for a long time without personal involvement save in so far as I am greatly interested in all that affects deeply the future of the national economy. With this matter of national economy all of us are concerned, and the relationship of the nuclear energy programme to the national economy is a crucial element of our debate. We in Britain are in no position to make wrong decisions for the longer-term future; our economy simply cannot afford to miss opportunities nor can it sustain wrong decisions. With no resources to squander we must be right. It is hoped the speakers cover most of the important aspects of the subject.

RELATIONSHIP WITH THE NATIONAL ECONOMY

If we were as wealthy and prosperous a country as we once were, in relative terms, the decision would not matter much. Any electrical power generated in the breeder side of an integrated nuclear programme might cost more than power obtained in a different fashion but this would not in itself seriously affect our industrial capability. However we are not in that happy position; indeed the situation is reversed.

I expect it will be shown in the course of the Meeting that an energy gap will exist within a very few decades. All of us know something about the effects of an energy gap. We have experienced a little of the possible repercussions because of OPEC policies: we pay several times as much for petroleum energy as we paid some years ago. The effects on the British economy have been extremely severe but we can still obtain petroleum. For that reason alone it is absolutely vital that a country, which is basically industrial in nature, does everything it possibly can do to avoid the impact of a real energy gap in the foreseeable future. My reading of most authorities on the subject satisfies me that there is no *proven* way of meeting the huge demands for energy that industrial society presents today without using fission power extensively. Estimates of how much that fission power must contribute within a few decades vary but most credible authorities put it relatively high. They estimate it must provide a large percentage of our total electrical energy needs, and indeed a comparatively large percentage of the whole of our energy consumption.

THE INCREASING DEMAND FOR ENERGY

At this point it is important to say a word or two about the rate of increase of energy consumption. Many proponents of alternative energy sources deny the fact that we need to consume an increasing amount of energy per capita per annum. They imply that man, in the advanced societies, is too concerned with adding to his material standards of living. It must be stressed that, even if we have no ambitions to improve our living standards but are content to maintain our present ones, our total energy consumption must continue to rise.

Recent estimates of the cost of conveying energy to the customer show that in the U.S. between 25 to 30% of the energy produced is used in winning, in transporting, and in distributing the energy, and it is clear that fraction will increase. The expenditure of effort, which implies expenditure of energy, increases with time as we try to exploit more inaccessible resources and North Sea oil serves as a good example. Much energy is expended in winning oil from under the sea, far more than is expended in exploiting it from much more accessible sources such as those found in the Middle East. Moreover, our raw materials of all kinds become slowly but inevitably less easily obtained and hence we expend more energy in making the same amount of these natural resources available. Many instances can be quoted. Iron ore, copper ore, manganese ore and many of the valuable raw materials become gradually less abundant and therefore more difficult to obtain, less easily refined and more energy-consuming in their preparation. In the field of foodstuffs the same is true. We need to feed more people while the population increases, but even to feed a static population in the same way costs us an increasing amount of energy. The manufacture of artificial fertilisers represents a good example of this need.

Without being at all greedy therefore man requires a steadily increasing amount of energy. To recycle materials is costly in terms of energy. This is a fact which is often obscured in the arguments advanced by those who are opposed to the increasingly costly and difficult means of obtaining all the energy essential to meeting human needs. It is a matter on which a deliberate choice must be made. Those of us in the less favourable climates of the world feel more energy is the proper answer. Scotland, a cold country, has become increasingly attractive as energy resources become available more abundantly.

ALTERNATIVE RESOURCES

Many of the proposed alternative energy resources will do little to meet our *major* needs. Thus wind and tides and even waves, while they may one day add a useful component, are incapable of meeting a large fraction, more than one-tenth say, of our total requirements. We are considering here the massive needs presently supplied through coal, petroleum, natural gas and nuclear energy. Even the last is as yet comparatively small compared with the others. Hydro-electric power, an admirable kind of resource, is small in comparison with the others. The only additional foreseeable major resources are in the conversion of sunlight by one means or another to a readily useful form of energy and controlled thermonuclear energy. While the work looks promising there is no certainty whatsoever that they will be found entirely practicable.

At the present moment controlled fusion has not been shown to be practicable in the laboratory. Even if something like a prototype machine could be built in the next decade, two or three further decades would be required in building and proving the early versions of energy-producing machines. Thus by the year 2005, assuming all went well, a small fraction of our energy requirements might be provided through fusion. We would not have reached a stage as far advanced as that of fission today. It would be another two or so decades at least before a substantial part of our total energy needs would come from thermonuclear plants. This is very late compared with the time that an energy gap will have developed.

Likewise solar energy conversion, while demonstrated on a minute laboratory scale, must be considered still in the research and development process; it would be foolish to believe that by solar energy conversion we could be adding the required large fraction of energy to our total pool in less than another fifty years. We must be clear that the *industrial* scale of major energy resources demands great investment and decades of manufacturing. To be at all realistic in handling programmes on the scale which is essential we must have continuity and that means steady sustained progress on the thermal and breeder reactors.

THE NUCLEAR PROGRAMME

If it is agreed that a large-scale fission programme of energy production is essential we then have to consider the nature of that programme. Many of the well-informed nuclear scientists and engineers have worked towards providing the best answer to that question for more than thirty years. Almost all of them – possibly all of them – advocate the utilisation of the breeder reactor. Even if that is agreed we still have problems in devising a well-integrated programme consisting of both breeder and thermal reactors.

There is no real discontinuity in proceeding to a programme of construction of breeder reactors. The thermal reactor is a major source of plutonium; it produces plutonium as the most important by-product within the reactor core. Indeed it is virtually the only way of providing ourselves with plutonium. Estimates show that hundreds of tons of plutonium have been made in thermal reactors in a number of countries. Thorium can be put to good use to provide an alternative integrated programme of nuclear power generation both by thermal reactors and breeder reactors. However very much less work has been done on the thorium possibilities and, as compared with uranium and plutonium knowledge, we are comparatively inexperienced.

THE PLUTONIUM QUESTION

A number of reports, including the Flowers' Royal Commission Report, have highlighted the problems that are presented in exploiting plutonium. It must be stressed however that there is nothing entirely new deriving from the introduction of a breeder reactor component of a total nuclear power generation scheme. It has been stressed that plutonium is potentially a dangerous radioactive material, though to experienced nuclear scientists and engineers it is comparatively benign. It must surely be emphasised that the breeder as such does not introduce an increase in the plutonium inventory. In fact the breeder element of a power programme could be used to reduce the total amount of plutonium in stock, including the plutonium that otherwise could simply be added to the weapon total. Indeed if 'peace' broke out we could use in power reactors of breeder form all the plutonium generated in the thermal reactors.

At the same time the hazards of natural radioactive materials, with which we have long been familiar, have been under-emphasised. I can recall personally when I would receive weekly one-tenth of a curie of radon from the Radium Institute in London. It came by mail in a relatively small lead container. I had to point out that I did not consider this practice to be a safe one, for the postman and others. What I want to stress here is that for most of this century we have been concerned with the completely safe handling of dangerous radioactive substances. The same properties that make them dangerous have of course allowed man to turn them to extremely useful and beneficial applications, as in medicine.

LIMITATION ON RADIOACTIVE WASTE

The Earth itself is not a safe place. Many of the major disasters are natural occurrences – for example, earthquakes, hurricanes and the like. Many fail to remember that there are hazards due to the natural radioactivity of the Earth's crust and this hazard is actually increased by burning coal. The Earth itself contains uranium and thorium and their families of radioactive substances, including radon gas. There are also large amounts of radioactive potassium constantly with us. It must surely be emphasised that in the process of using uranium in reactors the total number of radioactive atoms produced in a fission power programme cannot possibly exceed the number of radioactive atoms of natural uranium *by a factor of more than three*. The rate at which some of the man-made radioactive materials decay is very high which means they are very active while they exist but these high-activity substances mostly die while they are within the shielded reactor core. Some of the fission products left after re-processing prove awkward in their radioactive properties but in the final analysis there is no possibility of adding to the natural radioactivity of our Earth.

We must however consider one of the man-made materials – plutonium – in detail and take steps to safeguard its use. There is no reason to think we would fail. We have pointed out that in breeder reactors we would usefully consume it. In fact if one wishes to look far enough ahead, say several hundred years, at the end of that time we could finish with considerably less radioactive material on the surface of the Earth than exists in the natural state of that crust. We would in practice have burned a good deal of the uranium originally near the surface and we would have consumed much of the manufactured material such as plutonium. By that time we could hope to have found other means of liberating energy in the necessary huge quantities. So taking a really long-term view it can be argued that we never considerably exceed the naturally occurring amount of radiation around us and in the end we finish by reducing it in total. Many seem to imagine nuclear energy programmes result in the creation of massive amounts of dangerous radioactive waste but nothing could be further from the truth.

Before concluding this section I would like to stress that the nuclear industry started with a very great amount of knowledge of the nature of radiation, especially the penetrating radiation, and has therefore from the start been organised to minimise any risks arising. The results of this close attention to radiation hazards have resulted in an industry with a remarkably good safety record. It has been pointed out that if the New York Grand Central Station, constructed partly of granite, were a building for atomic energy use it would be condemned. This illustrates realistically the level of safety standards built into the nuclear industry.

A word or two about the safety of alternative resources or even other existing resources. No one would care to claim that North Sea oil exploitation has been easy – many lives have been lost and many are at stake. Likewise it is possible to believe that wave energy, when exploited, would create hazards for many. Perhaps solar energy is less hazardous but it is difficult to say till we know the nature of a major supply of solar energy.

THE TERRORISM MATTER

The breeder reactor is less complicated than a thermal reactor of the same output. Techniques of construction may be more demanding but it is more readily shielded and it would, for this and other reasons, present a really difficult barrier to the terrorist. It has nevertheless been suggested that in the plutonium age the action of the terrorist is potentially extremely serious. It is difficult to accept this argument; plutonium is a very valuable substance and treated as such. Moreover scientists could deliberately make it very hazardous to handle. It is safe to suppose that much less personally dangerous means of creating terror can be found. If informed scientists turn to terrorism then our society is in such a state that even the energy problem fades into insignificance. Few of us could conceive of such a situation arising.

CONCLUSION

I have touched on a number of the aspects of the subject which have featured in discussions and debates. We hope that the distinguished speakers who have gathered together here today will present us with a reasonably complete picture of the reactor power position and the rôle of the breeder. We trust the audience will be well informed and that the papers and discussions will help the nation to come to the right decisions.

The Birth of the Breeder

LORD HINTON OF BANKSIDE,
O.M., K.B.E., F.R.S.

Deputy Chairman, Electricity Supply Research Council

SUMMARY
The first electricity ever produced from nuclear fission was generated by a fast reactor at Idaho Falls in the late 1940s; since that time development has been continuous. In Britain it was realised that proven supplies of virgin uranium were limited so that it would be unrealistic to propose a nuclear power programme until there was certainty that the available uranium could be used economically by breeding fissionable material in fast reactors. Work on development and design of fast reactors was therefore started by Harwell and Risley in 1951.

The problems of design proved to be even more difficult than had been expected and (because defence projects had to be given priority) only a small team could be allocated to the work. A zero energy facility was built at Harwell and the small team of engineers at Risley started design of a 60 MW (thermal) experimental reactor. The core of this was only about 2 ft in diameter and 2 ft long and the removal of so large an amount of heat from this demanded a transfer coefficient far higher than any that had previously been continuously achieved in static plant; it was therefore decided to use a liquid metal coolant of which there was little previous experience. Neutron irradiation in the reactor core would make the coolant radioactive and a secondary liquid metal circuit was used to transfer the heat to water.

It was not known whether fast reactors would be as controllable as thermal reactors. The physicists feared that, if there was a failure in coolant supply, the fuel might melt and form a super-critical mass. It was therefore decided to build the reactor on a remote site and to contain it within a metal sphere; a suitable site was found near Thurso in the north of Scotland. Before construction started a public meeting was held in the Town Hall at Thurso at which a full and frank description of the work which was to be done and the extent of the environmental hazards was given. The reactor was commissioned in 1959; it was designed to give experience of the control of fast reactors, to give experience of the use of liquid metals as coolants and to provide a facility for the development of fuel elements suitable for industrial fast reactors. It was more than successful in all of these things.

There are few people today who remember that the first electricity produced from nuclear power was generated in a fast reactor. The Americans had built thermal reactors during the War to produce plutonium for atomic bombs; before long those reactors at Hanford became unsafe due mainly to dimensional changes in the graphite moderator. American interest in thermal reactors waned, they had large reserves of coal, oil and gas; their industry saw little chance of making money from thermal reactors and interest in them was kept alive only by the persistence of one man; Captain (now Admiral) Rickover. Rickover saw the importance of nuclear propulsion for submarines. The fast reactor did not ideally suit his requirements and he was responsible for the development of the Light Water Reactor which, with a vigorous sales drive, has now dominated the markets of the world. But while in those early years the industrial drive was for thermal reactors, physicists in America realised that uranium resources would be squandered if only thermal reactors were built and, by 1948, Walter Zinn at the Argonne Laboratory was working on the problems of the fast breeder reactor. It was his work that led to the construction of the EBR (the Experimental Breeder Reactor) at Idaho Falls and it was there that electricity was first generated from nuclear energy.

When the Atomic Energy organisation was set up in Britain its first task was to produce plutonium for defence purposes. We therefore had to build thermal reactors in which neutrons from the fission of uranium 235 are captured by uranium 238 with the ultimate formation of plutonium. While design and construction work on these first reactors was being done, the physicists at Harwell were already thinking about fast reactor design and when the Risley project office finished work on the Windscale Piles it was agreed that, after defence work, the fast reactor should be given priority. It was a wise decision; work on power-producing thermal reactors had already been done at Risley and at Harwell but the known resources of uranium were comparatively small and we felt that it was useless to aim at an industrial programme before we could be reasonably sure that uranium could be used economically and we knew that economical use was only possible with a balanced programme of fast and thermal reactors.

As was standard practice in those days, a joint Harwell/Risley design committee was formed in 1951 and it found that it was facing a problem even more difficult than it had expected. A Risley report written at that time and quoted by Margaret Gowing in her official history says,

> 'at first sight this fast reactor scheme appears unrealistic. On closer examination it appears fantastic'

the report goes on, with a philosophy typical of the Risley organisation, to say that the problem had to be solved and that

> 'Engineers solve the problems they have to'.

They did.

The Harwell physicists built a zero energy facility to provide nuclear physical data and Risley went ahead with only a small team to prepare schematic designs. In 1953 when it was necessary to start work on Calder Hall in order to meet additional defence demands, that fast reactor team shrank to about six engineers and draughtsmen and it was they who produced the feasibility study on which the final design was based.

No-one before had, continuously and over a long period, achieved, in static plant, the high rate of heat transfer that was required. About 60 MW of heat had to be removed from a cylinder 2 ft in diameter and 2 ft long. This required a heat transfer coefficient far higher than the best figure that had continuously been achieved in stationary industrial plant. For this and for other reasons we decided that we should have to use a liquid metal coolant; sodium was the best of all the alternative liquid metal coolants, but its one disadvantage was that it solidifies at about 100°C and we had enough problems without heat tracing every pipe and vessel. We decided therefore to use a sodium potassium alloy (known as NaK) which is liquid at room temperature. Neutron irradiation in the pile would make the liquid metal coolant radioactive so we had to introduce an inactive secondary liquid metal circuit within the biological shield to which the NaK would give up its heat. The liquid metal had to be circulated and no centrifugal pump had been developed in Britain which met our requirements. We therefore had to use electro-magnetic pumps which were available only in comparatively small sizes so that we needed a number of them. The use of a number of pumps (there were twenty-four in all) had the advantage of allowing us to use separate power sources for different pumps and this made total loss of coolant improbable. Because of the number of pumps, as well as for other reasons, we found it convenient to use a loop system for heat transfer from the primary to the secondary liquid metal circuit. I originally suggested the use of heat exchange circuits between these two streams of liquid metal which consisted of concentric tubes arranged helically round the inner periphery of the biological shield but the design office ultimately preferred to use concentric tubes in the form of a number of hairpins. Although the secondary NaK circuit was inactive it had to give up its heat to a water circuit and we decided that there must be double wall separation between the NaK and the water, we therefore cast the two pipes into a copper matrix to give a thermal bond.

It is the delayed neutrons that make control of thermal reactors easy but we were not sure whether we could count on these for control of a fast reactor. We arranged some of the fuel elements in clusters which though still remaining in the coolant, could be partly withdrawn from the core so as to reduce its effective size and activity while, in addition, boron control and shut-off rods were provided.

Although it was intended that plutonium should ultimately be used as the fuel, we knew that defence requirements were such that no plutonium would be available for use as fuel in our experimental fast reactor. We therefore used uranium enriched to $45C_0$, that is with 45 times as much fissionable U_{235} as in natural uranium. The high separation diffusion plant was being built at that time and we argued that we could get enough $45C_0$ material for our core from the medium separation stages before the high separation stages were in operation. The argument was, of course, spurious and I am sure that Lord Cherwell realised this and that with wisdom and foresight he connived at the diversion of sufficient $45C_0$ material for the first core.

In those days the technique of manufacturing oxide fuel had not been developed and so we had to use metallic uranium which has a comparatively low melting point. With this low melting point fuel there was a possibility that loss of coolant could cause the fuel to melt and the physicists predicted that if this happened and the fuel ran down into the bottom of the vessel and formed a super-critical mass we might have an explosion with the force of half a ton of TNT. My diary records that on August 5th, 1953

> 'I sent Kendal, Hill (who is here today as Sir John) and Butler down to Harwell to try to knock some sense into the safety distances for the Fast Reactor'.

Argument about safety went on for many months; Lord Cherwell laughed at the idea that a minor nuclear explosion was possible but we decided that in view of all the uncertainties that existed, we would put the whole assembly inside a steel sphere.

Because of the unknowns we had, from the outset, planned to build the reactor on a remote site. Kendal, Owen and I had examined the Galloway and South Ayrshire coasts without finding anything that was suitable. We decided that we had to go further afield and surveyed the coasts of Sutherland and Caithness and there we found what we knew to be the best site at Dounreay. But why, if we were giving the reactor containment, were we putting it on a remote site? This could only be logical if we assumed that the sphere was not absolutely free from leaks. So we assumed, generously, that there would be 1% leakage from the sphere and, dividing the country around the sites into sectors, we counted the number of houses in each sector and calculated the number of inhabitants. To our dismay this showed that the site did not comply with the safety distances specified by the health physicists. That was easily put right; with the assumption of a 99% containment the site was unsatisfactory so we assumed, more realistically, a 99·9% containment and by doing this we established the fact that the site was perfect. I mention this simply to show the change in philosophy in the years that have passed since 1953 – we knew that we had found the right site for the reactor and we were quite prepared to adjust what were only guessed figures to support a choice that we knew from experienced judgement was right. Today a collection of guessed figures would be fed into a computer and people would image that whatever answer came out was bound to be right. Some day I hope that it will, to the great advantage of engineering, be realised that mathematics is its servant and not its master.

It was not until the end of 1954 that we were given financial sanction to build the experimental reactor on the Dounreay site and I decided that it was right to give a full explanation of what we were doing to the people of Thurso in a meeting at the Town Hall. My wife and I went by sleeper to Inverness; it was her first visit to Scotland and she had never travelled by sleeper before. I promised her a comfortable night and told her that they put a restaurant car on to the train at Perth and that we should get a breakfast to be remembered. Unfortunately I made what I think of as my only bad mistake in connexion with Dounreay – I arranged to travel on the night after Hogmanay. There is no need for me to tell an audience in Glasgow that breakfast was a disappointment.

In the Town Hall on the next evening I gave the lecture which I had regularly given in our induction courses at Risley and which explained the simple theory of atomic energy – that lecture was later given on the Overseas Service of the BBC and was published by them in a booklet that was translated into 26 languages. I followed the explanation by an evaluation of the hazards that could arise from the fast reactor and a description of the precautions that had been taken – the script covers four sheets of foolscap but there are three paragraphs that I should like to quote today; in those paragraphs I said,

> 'Let me make it quite clear that I am not going to claim that there are no risks associated with the development of Atomic Energy. Every human activity involves a certain measure of risk and we can only evaluate any one risk by comparing it with the other risks to which we are subjected.
>
> 'By far the greatest risk that we run is that of dying from natural causes. If we consider the group of people whose ages lie between 10 and 45 we find that 150 out of every 100,000 die each year. Nearly 40 of these die deaths of violence; they are electrocuted in their homes, they are the victims of road accidents, they fall downstairs or they commit suicide. The rest of the 150 die from disease or illness.
>
> 'Every man below 45 years of age runs this risk. If he works in the dangerous industries such as quarrying or coal mining his risk of death in any one year is about doubled. In the "safe" industries such as engineering or the electrical industry his chance of death is brought up from 150 to 160 – an increase of 7%. In the atomic energy industry we find that the corresponding increase is only 2%'.

I went on to explain the nature of the risks and the precautions that were taken.

On December 22nd 1976 the Report of the Royal Commission on 'Nuclear Power and the Environment' was debated in the House of Lords. In that debate Viscount Thurso said,

> 'I remember a talk that was given in the Town Hall in Thurso – I thought that I could get in quite easily but in fact I had to listen from outside the main door because the Hall was packed – that talk was given by Sir Christopher, now Lord Hinton. In that one talk – it was longer than his speech in your Lordships House today – he explained in clear terms to the people of Thurso what the Atomic Energy Authority hoped to do in their midst and how it would affect them. I think, 21 years later that I am able to say that he did not mislead them. My experience is that we were told the truth, what Dounreay would be like and how it would affect our lives. I have watched the reactors in construction and in running and I have tremendous confidence, built from knowing the people who work in the industry and from knowing their sense of responsibility to the human race as well as to their profession'.

It was I who gave those first assurances to the people of Thurso. I was responsible for that first fast reactor at Dounreay; I have not been responsible for the prototype fast reactor which is now in operation but I am reasonably well informed about it because I have reported on it for the Government or for the Atomic Energy Authority on three separate occasions. Some rectifiable mistakes have been made, not in the design of the reactor but in the design of the ancillary plant; but I am prepared to say the same things about the prototype fast reactor in 1977 that I said about the experimental fast reactor when in 1955 I spoke at the meeting in the Town Hall at Thurso. The next step forward is to build a full-scale industrial fast reactor. Provided the design of that reactor is done by competent engineers working in an organisation which is not complex and in which the ultimate responsibility is clearly defined, I am sure that I should be able to say the same things about it as I said about the First Dounreay Fast Reactor in 1955 and the things that I am saying today about the Prototype Fast Reactor.

The Energy Gap and the Fast Reactor

SIR JOHN HILL

Chairman, Atomic Energy Authority

SUMMARY

The fast reactor, with its immense potential, will be the most obvious source of the energy required worldwide in 25 years' time, and will become an essential component of the world energy scene. Toward the end of the century, oil and gas will be in short supply. The conversion of the world's economy from oil and gas back to a coal economy is theoretically possible. The real issue, however, is not whether there is sufficient coal in the world, but whether energy-hungry parts of the world are going to be able to get access to the quantities they will need at prices they can afford. Thermal reactors can make a significant contribution to the world energy scene, but they cannot transform it. The fast reactor, with its 50 times better uranium utilisation, can.

In Britain, the results of the many experiments performed, the operation of the Dounreay Fast Reactor for the past 18 years and the first year's operation of the larger Prototype Fast Reactor have all been very encouraging, in that they demonstrated that the performance corresponded well with predictions, breeding is possible, and the system is exceptionally stable in operation.

The next step in fast reactor engineering is to build a full-scale fast reactor power station. There would seem to be little reason to expect more trouble than could reasonably be expected in constructing any large project of this general nature.

However, from an engineering point of view continuity of experience is required. If a decision to build a commercial fast reactor were taken today there would be a 14-year gap between starting this and the start of the Prototype Fast Reactor. This is already much too long.

From an environmental standpoint we have to demonstrate that we can manufacture and reprocess fast reactor fuel for a substantial programme in a way that does not lead to pollution of the environment, and that plutonium-containing fuel can be transported in the quantities required in safety and in a way that does not attract terrorists or require a private army to ensure its security. Finally, we have to find a way to allow many countries to obtain the energy they need from fast reactors, without leading to the proliferation of nuclear weapons or weapons capability.

I base the case for the fast reactor on a very simple argument. The world is going to be very short of energy in 25 years' time and mankind will be exploiting every source of energy at his disposal. The fast reactor with its immense potential will be the most obvious source of the energy we require and will become an essential component of the world energy scene.

This view is challenged by all the detractors of nuclear power who have made the fast reactor and the plutonium fuel it uses the principal target of their attacks. To establish my case I have to demonstrate three things:

(i) that there is going to be a major world energy shortage in about 25 years' time;

(ii) that the fast reactor is a practicable technology which can make a material contribution to world energy requirements;

(iii) that the industrial scale use of the fast reactor need not lead to the proliferation of nuclear weapons, pollution of the environment or precipitate other unacceptable side effects.

I will attempt to establish these three foundations to my case.

From general principles it can be seen why the world is going to need more energy in the future. First there is an almost direct relationship between gross national product and energy consumption. We can certainly save energy by conservation, but the linear relationship covers such a vast span, that the general philosophy that material growth requires more energy cannot be denied. Even if the wealthy countries are prepared to limit further growth of affluence, the poorer countries, where real deprivation exists, and where more than half the world's population lives, want growth more than anything else.

Secondly, world population is going to expand very substantially over the next fifty years. One has only to look at the age distribution of the population of the world to see that this is inevitable. In Britain and many advanced countries we have a stable population distribution with there being nearly as many people of 60 as there are of 16. In the poorer countries the population distribution looks like a triangle with young people at the base. Even if the whole world were to enforce from today rules which limited families to the same size as we have in Britain there would still be an enormous increase in population before the age distribution reached stability. This increased population is going to require an equivalent increase in energy, even with no increase in the standard of living.

If the world is to limit its energy consumption to the existing level then either the richer countries have to use very much less than they do today or the poor countries must get poorer and poorer year by year.

At the present time nearly three-quarters of the world's effective energy input comes from oil and gas. There is enough to last at current rates of concumption for about 40 years, or with some growth perhaps 30 years – not very long, only one generation. But the situation is more serious than this because the shortage comes not when we have burnt the last drop of oil or the last puff of gas, but when production can no longer expand at the rate of increasing demand. This point of time is about 15 years away, even assuming no political restriction on the oil companies' ability to extract and distribute hydrocarbons on a world scale. We will indeed be fortunate if there is no interruption to world oil and gas supplies over this period of time.

There is certainly very much more coal available in the earth than oil or gas, perhaps ten times as much. Coal is less convenient than oil or gas, but it can be used perfectly well as a heat source for electricity generation, many factory processes and space heating. It can also be used to manufacture substitute natural gas and oil and petrol. The conversion of the world's economy from an oil/gas economy back to a coal economy is certainly theoretically possible. The real issue is not whether there is sufficient coal in the world, but whether the energy-hungry parts of the world are going to be able to get access to the quantities they will need at prices that they can afford.

Of the world's total reserves of coal, over 50% lie in Siberia, in deep seams. It is an area remote from the energy consuming industries of the world, it is remote from the sea. It is an area of extremely harsh climatic conditions, an area of very low population and, from the point of view of coal handling, an area of non-existent communications.

The second major reserve of coal is in the United States of America. Not in the well-known coal fields of the east coast and Appalachian Mountains, but in the Mid-West. Again very large distances from the energy consuming areas.

Coal is very much more difficult to handle and transport than oil. Certainly large bulk carriers can move the coal over large distances at sea, but transport over substantial distances of land is more difficult. A train carrying 2000 tons of coal represents a very heavy goods train, over twice the weight of the coal trains hauled by British Rail. Such trains passing every quarter of an hour, day and night, every day of the year would carry about 70 million tons of coal per annum. Even on this assumption of what a railway could carry, a double-track railway from Siberia to the U.K., exclusively devoted to coal traffic would not sustain the Central Electricity Generating Board, never mind the rest of British industry.

I draw this extreme example to illustrate that we must not in our thinking take figures of world reserves and assume that they can be made available to us in any reasonable time. Coal will certainly become a fuel of international trade, but we must recognise that it will be expensive and the sheer problem of producing vast quantities from areas new to this activity and the difficulties of handling and transport will mean that even in 25 years' time, the majority of the world's coal will be produced fairly close to the point of consumption. This is, in a way, a reversion to the energy situation of 40 years ago, but now the problem is greater in that energy consumption has quadrupled.

What is daunting is the scale of the operations required. What are the implications of Japan importing 800 million tons of coal a year, or the United States mining 4000 million tons, or even Britain having to mine 400 million tons? These figures cannot of course be reached on anything like the timescale we are discussing, and the industrialised world is going to be heavily dependent upon imported oil until well into the next century. But who is going to be a willing seller at a time of declining production and in clear sight of the complete exhaustion of oil supplies? Even if allowance is made for the contribution which can be expected from the alternative energy sources such as the wind, the waves and the various solar energy systems, this does not significantly alter the overall picture which I have outlined of a world in which a steadily increasing demand for energy has to be met by reliance in the main on rapidly dwindling stocks of oil and gas, and on the supply of coal which is going to be increasingly difficult to exploit economically as a world fuel.

Reasonably assured uranium supplies would, if consumed in thermal reactors, produce somewhat less total energy than the world's oil supplies. These reactors can therefore make a significant contribution to the world energy scene but they cannot transform the position. The fast reactor, with its fifty times better uranium utilisation can do just this. What then are the real limitations to the expansion of the fast reactor system? Let us consider the subjects one at a time.

1. PHYSICS

Most of the physics problems of the fast reactor are now understood. We started to investigate these problems in 1954 when we built the first very low power plutonium-fuelled fast reactor. When it was decided to build the experimental fast reactor at Dounreay a low energy mock-up of the core was built at Harwell and the fuel was later transferred to Dounreay.

The results of these many experiments, the many years of operation of the Dounreay Fast Reactor and the first year's operation of the Prototype Fast Reactor, have all been very encouraging in that they demonstrated that the performance corresponded well with predictions, breeding was possible even allowing a generous margin for engineering structural materials and the system was exceptionally stable in operation.

2. THE SODIUM COOLANT

The coolant selected was liquid sodium. This decision was based on straightforward physics considerations. Hydrogen containing liquids were excluded because of the degradation of the neutron spectrum and gases did not appear to have a sufficient heat transfer capacity to achieve the desired fuel ratings. Time has confirmed the correctness of this courageous decision. But when the decision was taken the use of liquid sodium as a heat transfer medium was virtually untried.

At first we found liquid sodium a difficult material to work with. We got blockages with sodium oxide and hydroxide and we had corrosion problems. But it rapidly became clear that the problems were not related to liquid sodium but to the impurities in the sodium and as soon as we developed efficient traps for removing the oxide and other impurities, the problem disappeared. Now when you hear a fast reactor operator complaining about the difficult and corrosive liquid he has to deal with, he is talking not about the sodium but the water. This is literally true – ask any fast reactor man in the audience.

Another major area of concern was the problem of a sodium/water reaction if, or rather when, faults develop in the steam generators. For the experimental reactor at Dounreay we avoided this problem by putting the sodium and water in separate pipes and bonding the pipes together with copper.

For the Prototype Fast Reactor we faced this problem squarely and set ourselves the task of building single skin steam generators and superheaters to provide steam at 160 bar and 565°C, i.e. the steam conditions of contemporary fossil fuelled plant. We nearly succeeded first time and it is indeed fortunate that we encountered the difficulties when we did and not at the next stage. Leaks were found in about 10 welds out of the 10,000. They were all detected rapidly at a stage where they were so small that they were extremely difficult to locate. They would probably have plugged themselves in a more normal environment. The austenitic steels that we used in the superheaters and re-heaters and which is necessary at temperatures of 565°C we now recognise as being too vulnerable to caustic stress corrosion to be acceptable for single-skinned sodium to steam heat exchangers. We have decided to lower temperatures slightly and revert to ferritic steel which has proved entirely satisfactory for the evaporators. In view of the difficulty in locating very small leaks in the welds we have modified the design to avoid welds being located in positions where there is steam on one side and sodium on the other. With these relatively simple changes we believe we can in the future avoid the problems encountered in the earlier design of steam generator.

3. THE FUEL

A problem that worried us for a long time was fuel and fuel performance. Experience, however, could hardly have been better. Replicas of PFR fuel have been irradiated extensively in DFR for many years and the first charge of PFR fuel has operated to date with no failures. Fast reactor fuel is in fact the most tolerant nuclear fuel that has been developed for any reactor. Experiments have been carried out in the Dounreay reactor where fuel has been starved of coolant to the extent that the sodium was boiling in the sub-assembly without any damage taking place. In other experiments fuel has been pushed to the point of clad melting without incident.

4. ENGINEERING

Fast reactors require high quality engineering, but this is available and the actual reactors of the two fast reactor power stations we have built have given hardly any trouble at all. Most of the troubles we have encountered with the Prototype Fast Reactor have been on the conventional steam side where standards had not been set at the very high levels adopted for the reactor.

The next step in Fast Reactor engineering is to build a full scale fast reactor power station. It is proposed that the key components such as fuel, control rods, heat exchangers, should be made approximately the same size as in the prototype reactor but employing more units in parallel. The sodium pumps would be about double the size at present in use. There would seem to be little reason to expect more trouble than could reasonably be expected in constructing any large project of this general nature.

But I would be less than credible if I were to pretend that there is nothing more to learn and that all problems are solved. What then is it that we are hoping to learn or demonstrate over this final phase?

From a physics point of view we have to do more work analysing what can happen under highly improbable but extreme core accident situations. We have to consider if changes in core design will make it even more tolerant of accident conditions. We have to obtain more experimental confirmation that volatiles in the fuel, whether they be fission products or additions to the fuel during manufacture, will limit the maximum possible energy release to that which can be contained with certainty within the structure. This is a pre-requisite of the Nuclear Inspectorate being prepared to authorise near urban siting and which is in turn an almost essential requirement of a large programme in a country such as ours.

From the point of view of the sodium coolant we need to get more experience on the effect of thermal shock on the reactor structure and how any detrimental effects can be engineered away by appropriate design or thermal insulation.

On fuel, we need to do more work on the fabrication processes to enable a substantially larger throughput of plutonium containing fuel to be achieved without increase in the production of residues and contaminated waste or an increase in operator exposure. Second generation fuel fabrication plants are already at the prototype stage to achieve these objectives. What we are proposing to do is to use wet processes to a greater extent than previously and to eliminate the use of powders or any dust-producing stages in the manufacture of plutonium fuel. There are several routes by which this can be achieved. What we have to evaluate is which is the best.

From an engineering point of view, we require continuity of experience. If a decision was taken today there would be a 14-year gap between starting the Prototype Fast Reactor and CFR 1, the first full-scale demonstration plant. This is already much too long.

Finally we have from an environmental standpoint to demonstrate that we can manufacture and reprocess fast reactor fuel for a substantial programme in a way that does not lead to pollution of the environment. We have to demonstrate that plutonium containing fuel can be transported in the quantities required in safety and in a way that does not attract terrorists or require a private army to ensure its security. We have, finally, in the long term, to find a way to allow the many countries of the world to obtain the energy they need from fast reactors without leading to the proliferation of nuclear weapons or nuclear weapon capability.

To achieve the manufacture and reprocessing of substantial quantities of fast reactor fuel without pollution to the environment is simply a matter of care, good design of plant and money. We are continually up-grading our processes with this objective. The new fuel fabrication processes are now, as I have just described, at the prototype stage and have the objective not of making better fuel, but to avoid producing residues and waste. New incinerators are being developed to recover plutonium from contaminated waste, not for the value of the plutonium, but to avoid the problem of disposal. New evaporators and decontamination plants are being built to reduce the level of activity discharged to the sea.

We can achieve whatever level of containment we require or are directed to meet. My only plea is that the rules should continue to be set at sensible levels to prevent the unnecessary expenditure of what is really public money which could be used better in other ways.

The transport of plutonium fuel for a fast reactor programme is a non-problem. We need have no fear of terrorists attempting to steal fuel in transit and we need no armed guards to protect it. One way of avoiding this problem is to build the fuel reprocessing plant and refabrication plant alongside the fast reactor or reactors in which case the transport problem disappears. The other solution is to give the fuel a light irradiation in a small reactor built alongside the fuel fabrication plant in the central reprocessing facility. If this were done all fuel movements would be carried out in an active condition in heavy shielded containers and a terrorist could gain no access to the fuel.

In my view there can be no question that the fast reactor can provide this country with the electricity it requires in the future. It can do this more economically than by burning fossil fuels which might then be used for other purposes for which nuclear power is not suitable.

The fast reactor can do this with as high a degree of safety as any other way of getting the energy we require. It can do it with less impact upon the environment than wind power, wave power or solar power. All we need is the will to get on with the work.

The Dounreay Project

C. W. BLUMFIELD

Director, Dounreay Nuclear Power Development Establishment

SUMMARY

The paper briefly refers to the history of the Dounreay Experimental Reactor Establishment and the 18 years of safe operation of the Dounreay experimental liquid-metal-cooled fast reactor (DFR). Reference is also made to the successful commissioning of the 250 MW(e) Prototype Fast Reactor (PFR). This long history of the use of fast reactor technology has shown that although fast reactors are different in concept to thermal reactors they are as easily operated, if not more so.

The experience gained from the scientific research into materials suitable for fast reactors has provided a comprehensive library of information for the design of the next phase of fast reactors.

Fuel assemblies have been operated at full power with flow restricted to produce coolant boiling conditions. These have shown that the high heat fluxes of fast reactors are adequately tamed by the liquid metal coolant.

Experience in commissioning the PFR has enabled the designers of the demonstration Commercial Fast Reactor to progressively deal with important issues. In particular, the boiler design, which was always considered to be a difficult technological issue, has evolved to eliminate the small water leaks that have occurred in the PFR boilers. This will be tested in the PFR well in advance of its use in the demonstration Commercial Fast Reactor.

Reprocessing of PFR fuel at Dounreay will commence in about a year's time. This has involved reconstructing a plant which was used for reprocessing of DFR fuel. The decommissioning of the plant to the colloquial green field state allowed the modifications to be carried out expeditiously. The new plant has been designed on the knowledge produced in experiments to determine the most efficient and safe method of separation and waste management.

INTRODUCTION

The Dounreay Experimental Reactor Establishment (DERE) was built mainly to investigate operational aspects of liquid-metal-cooled fast reactors including the fuel cycle.

This led to the construction of two fast reactors, the Dounreay Fast Reactor (DFR) together with fuel fabrication and reprocessing equipment, and later the Prototype Fast Reactor (PFR) and its fuel reprocessing plant.

Associated with these are metallurgical and chemical R&D facilities to investigate materials performance.

Just over 2000 people are employed at Dounreay and about half were recruited from the locality.

BRIEF HISTORY OF ORIGIN OF THE DERE SITE

Lord Hinton in his paper has dealt with the early history of Dounreay. For those who do not know the site the photograph (Fig. 1) shows the blend of local agriculture with advanced technology. It is interesting to reflect on the development of the reaper binder which was mainly by experience. At Dounreay we have developed fast reactors by lengthy experience – over eighteen years – and of course much detailed scientific research.

The experimental Dounreay fast reactor (DFR) has safely operated since 1959 and together with its fuel reprocessing and fabrication plant, materials and chemistry development and operational reactor physics development has placed the U.K. in the forefront of world fast reactor technology. Furthermore it has shown that large fast reactors can be safely operated to an exacting scientific programme by employees without previous experience – people who were employed in agriculture and fisheries.

BENEFITS FROM DFR

Shortly after initial operation sufficient data were available to show that the DFR was easy to operate – in fact easier than thermal reactors. It had a negative power coefficient and because the margin to coolant boiling was large, the power could be allowed to rise by a factor of two above normal operating level for a lengthy period without significant damage to the fuel.

The DFR was to a large extent based on an extension to the technology for metallic fuel developed for experimental thermal reactors. Shortly after commencement of operation it was found that the fuel burn-up would be inadequate for an economic fuel cycle for future fast reactors. The rôle was then changed to that of testing fast reactor materials, another far-sighted decison. The modification to the reactor core meant that a concerted attack on the effect of liquid metal and fast neutrons on the physical characteristics of fuel and other materials could be made. At any given time there have been up to 100 experiments in the reactor. The graph (Fig. 2) shows the number of experiments loaded in the reactor over its lifetime.

Important aspects of this work are the endorsement of the performance of plutonium-based oxide fuel which it was anticipated would be the initial type fuel for commercial fast reactors because it used much of the technology evolved for the later power station type thermal reactors. The DFR showed at an early date that with such fuel a burn-up of heavy atoms of 7·5% (about 75,000 MW days/ton of heavy atoms) could be achieved compared with the maximum of 30,000 MW day/ton for thermal reactors. Subsequently some fuel has been taken to over 20% burn-up without failure and without adverse effect on disassembly of fuel for reprocessing. The long term implications of this work on the economics of fast reactors is far reaching, as obviously the larger the burn-up achieved the cheaper the fuel costs and the less the reprocessing waste. This work has also had a world wide effect – all countries are now pursuing a similar design of fuel assembly and the mixed plutonium oxide/uranium oxide route.

Fig. 1. General view of Dounreay site.

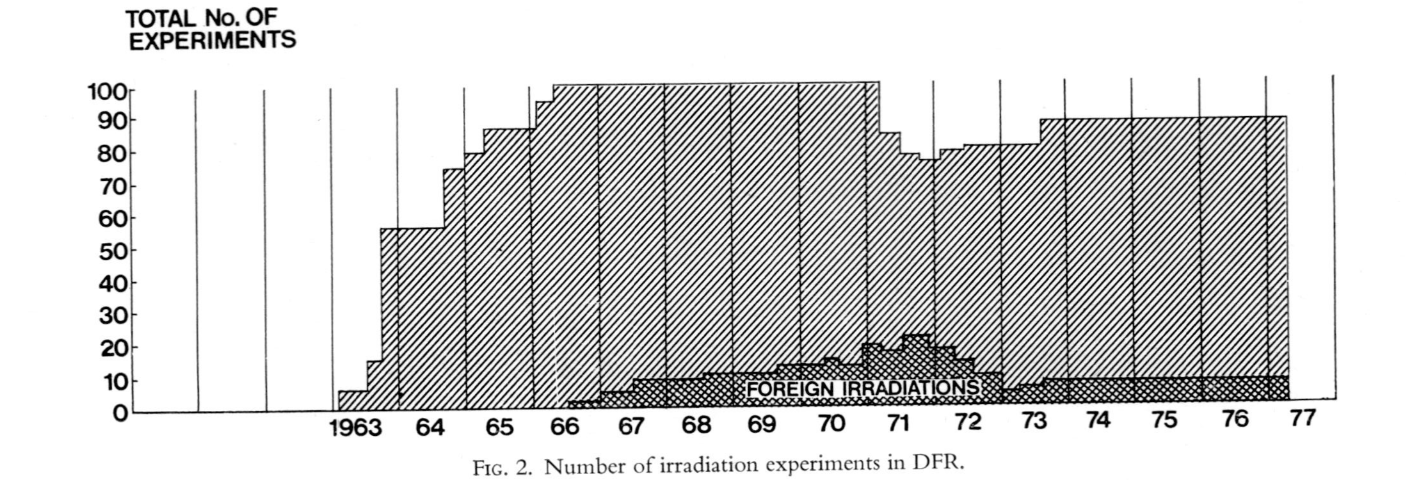

FIG. 2. Number of irradiation experiments in DFR.

Important physical properties of fuel cladding have also been pursued. The ductility, creep properties, corrosion resistance, mass transfer, crystal structure and ultimate strength are but a few of the properties which have been measured on innumerable samples over a wide range of temperature and irradiation conditions. This has given a vast library of design information. Some materials have been irradiated to the extent that each atom in the materials has on average been displaced 80 times. The surprising thing is, that even so, the materials changed very little and the effects can be readily dealt with by design. The graph (Fig. 3) shows how diametral strain in preferred fuel cladding material varies along the length of the fuel pin and at various burn-ups. It is interesting to note that only 4% strain occurs after 20% burn-up of the fuel – a very acceptable performance.

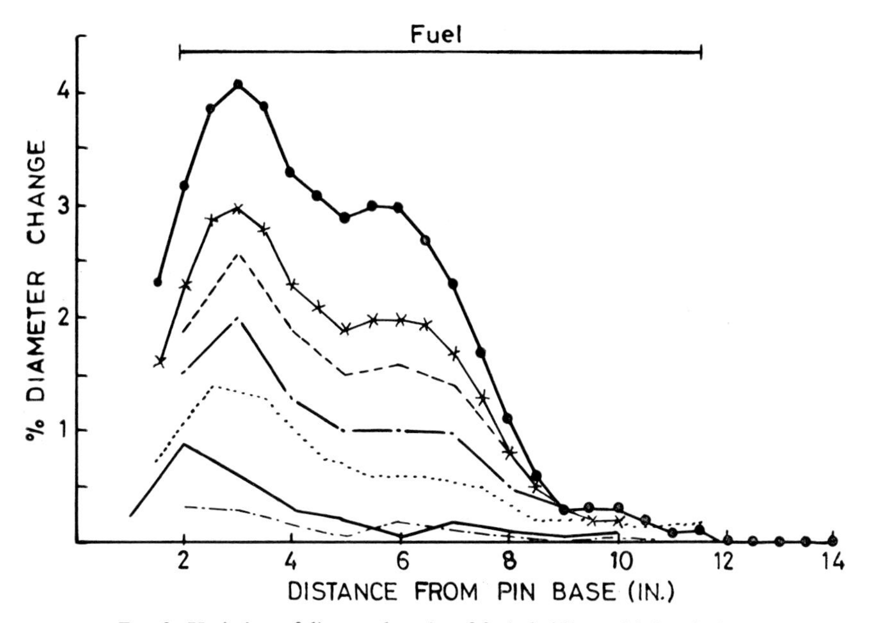

FIG. 3. Variation of diametral strain of fuel cladding with irradiation.

A major discovery took place in 1966. Fast neutron irradiation of materials produces vacancies and interstitial atoms within the lattice. Within a certain range of temperatures the vacancies migrate to nucleation points to form voids. The photo-micrograph (Fig. 4) shows at a magnification of 14,000×

the crystal structure of M316 stainless steel at an irradiation level of 42 displacements per atom. The void forms are obvious and many of them are greater in diameter than 1 micron. This causes swelling of the material. The graph (Fig. 5) shows the temperature and neutron flux variation along typical fuel pins and the associated voids in the fuel cladding at various points. At temperatures around half the melting point of the material voids are at a maximum. At temperatures slightly below and above there are little or no voids. In fuel assemblies situated where a gradient in neutron flux intensity occurs, fuel assemblies could bow. Some materials are orders of magnitude better than others in resisting void growth and deformation. Further fundamental research is being carried out into the effects of impurities. Also operational procedures to rotate fuel assemblies at intervals show that this will solve the problem but less elegantly than a fully-resistant material.

It has been proved that continued operation of fuel for over $1\frac{1}{2}\%$ burn-up after cladding has failed does not adversely affect the geometry of the fuel or the coolant channels. The photograph (Fig. 6) shows one of the large number of fuel pins taken to $1\frac{1}{2}\%$ burn-up after failure. The fuel material remains in position and the geometrical distortion is small and does not adversely affect heat transfer to the coolant. This is both important to safety and operation. The benign failure mechanism precludes the need for immediate action as soon as a clad failure is detected.

Many experiments in which the coolant has been deliberately throttled to make the coolant boil in fuel assemblies have shown that even over a period of hours the fuel assembly geometry remains acceptable. Acoustic detectors in DFR were developed to automatically trip the reactor should boiling occur. Another important safety feature is that the flow through a sub-assembly has to be reduced by two-thirds before coolant boiling occurs and this requires a 90% blockage area across the fuel assembly. The coolant in a fuel assembly is normally about 500°C below boiling point and in by-passing the blockage provides a sink into which any locally produced vapour rapidly condenses thereby giving stable flow conditions.

Numerous fuel pins have been operated at extremely low flow and clad and fuel caused to melt. The molten material migrated a short distance and froze. There has been no evidence of a rapid transfer of heat energy from molten fuel to coolant which theoretically could produce a pulse of mechanical energy. These were important experiments relative to safety arguments about possible progressive damage to other fuel assemblies.

In fact the high heat fluxes of fast reactors appear to be more than adequately tamed by the liquid metal coolant. It is not an over-emphasis to say that the combination of a high-conductivity low-pressure coolant with a boiling point nearly a factor of two above normal operating temperature is of major safety importance.

The long experience of DFR fuel reprocessing has shown that highly radio-active highly enriched fuels can be processed safely by solvent extraction. The original concern about adverse effects of irradiation on the solvent has proved to be largely unfounded. There is very little degradation and the efficiency of extraction is not impaired even when the solvent is recycled

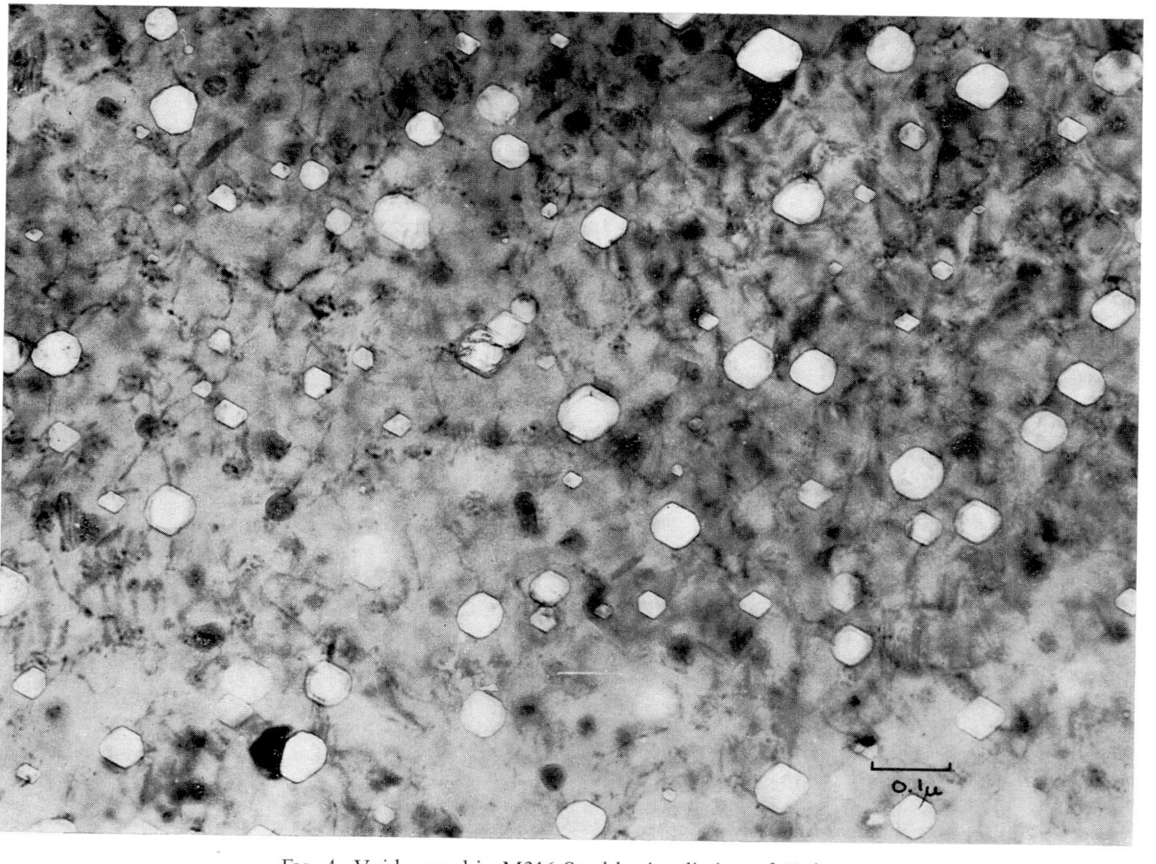

FIG. 4. Void caused in M316 Steel by irradiation of 42 dpa.

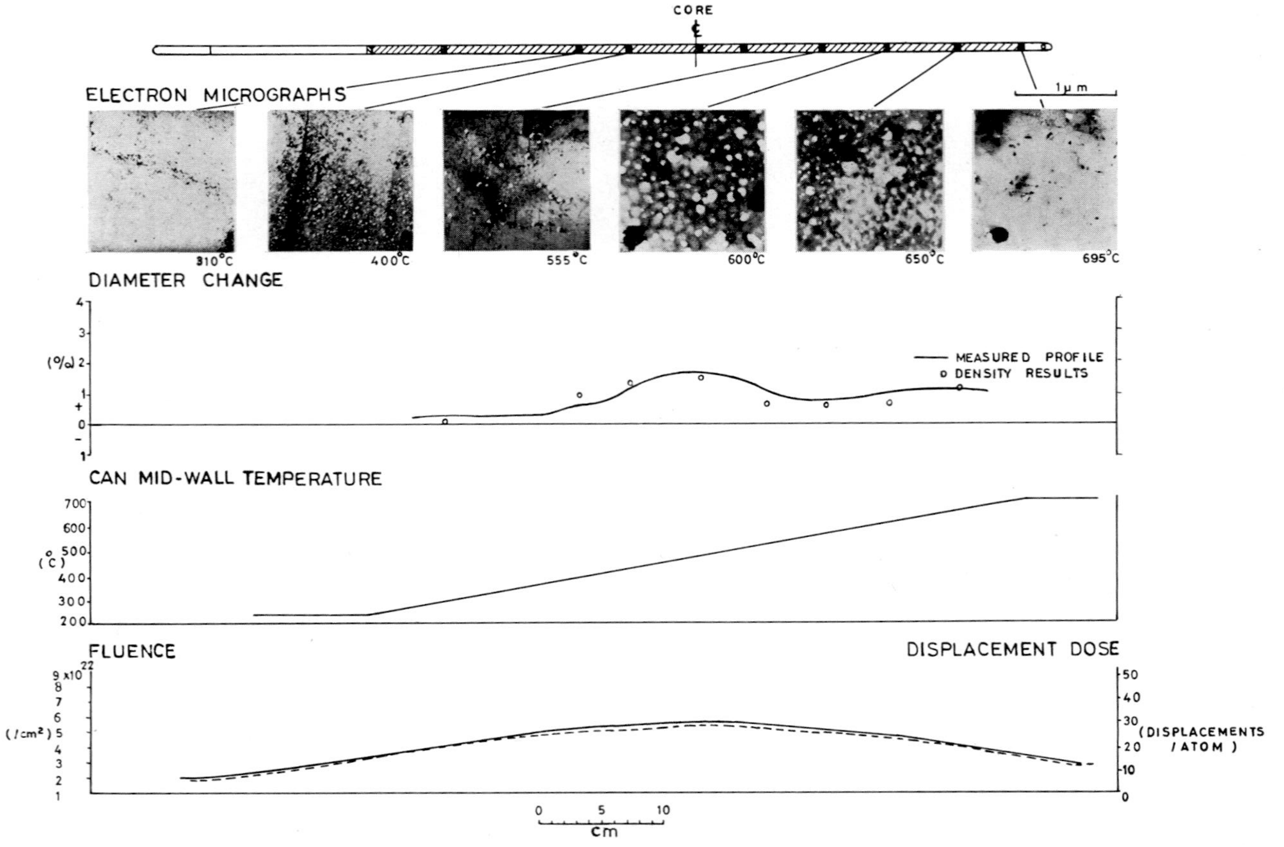

Fig. 5. Variation of diameter and voidage in FV 548 steel cladding related to temperature and neutron dose.

thousands of times. The two cycles of solvent extraction give a fission product decontamination factor of nearly 10^7. Since reprocessing commenced in 1959, 10^8 curies of activity have been extracted and stored without problems. Recently the plant has been decontaminated and is being refurbished for reprocessing PFR fuel. This important stage in its development has shown that such equipment can be decontaminated to the level required for continuous access and for a return to 'green field' use.

The DFR was closed down on 23 March 1977 at which time the fast reactor materials experimental work was taken over by the Prototype Fast Reactor. Over a period of a few years the reactor will be decommissioned to the stage where it would be possible for it to become a fast reactor museum which could be visited by the public.

THE PROTOTYPE FAST REACTOR (PFR)

The layout of the PFR buildings relative to the DFR is shown in Fig. 7. A stylised flow diagram of the reactor, sodium circuits, steam plant and turbo-alternator is shown in Fig. 8. The system was designed to allow the reactor to be safely sited at Winfrith in Southern England or at Dounreay in Northern Scotland. Each of these sites had expertise in aspects of fast reactors. Dounreay was selected by the Government on sociological, not safety grounds.

PFR is a true prototype. It was designed to produce 600 MW of heat and 250 MW of electricity. In agreement with the Electricity Boards and the Nuclear Industry full-sized fuel assemblies and boilers for commercial fast reactors (CFR) were the basis of the design. Also other components for CFR would only have to be extrapolated by a factor of 2 to $2\frac{1}{2}$ and experimental work indicates that these are readily achievable. The temperatures are slightly higher than for the CFR because they were based on the then requirements for the 300 MW turbo-alternators being purchased by the Electricity Boards. In addition, the PFR is supported by a total fuel cycle, i.e. fabrication, irradiation, reprocessing and refabrication etc.

The design of such a pool type reactor lends itself to disposal at the end of its life. All of the core and breeder components are removed for reprocessing, the shield units are readily removable and the surrounding components are virtually inactive.

The major points that have arisen since electrical power was first produced in late 1974 are related to the small leaks in the tube-to-tube plate welds of sodium-heated main boilers, problems with conventional steam plant and the high availability of the reactor (about 90% over the last 9 months).

Experimental work at DERE over a long period had shown that the boilers could be adequately designed to safely contain a large water/sodium reaction in the event of water-tube failures. For small leaks a very sensitive leak detection system based on diffusion of hydrogen through a nickel membrane both in sodium and in inert gas above the sodium was developed. The PFR boilers, a typical cross-section of which is shown in Fig. 9, were welded to exacting specifications, radiographed, crack detected, hydraulically tested, helium leak tested and passed for installation. They were again helium leak tested at

FIG. 6. Extent of fuel pin cladding damage after 50 days' irradiation following initial failure.

Fig. 7. General view of PFR buildings.

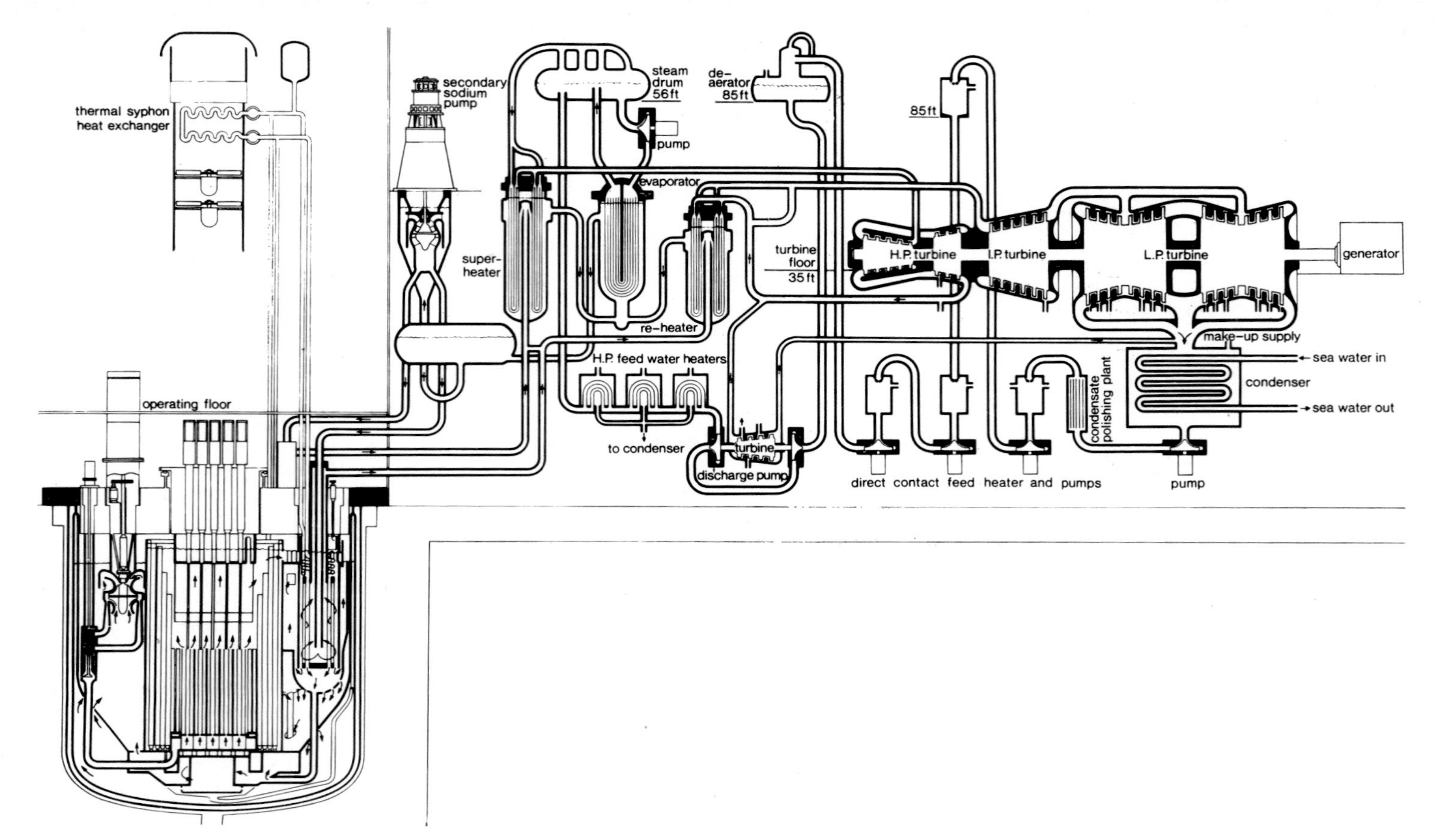

FIG. 8. Schematic diagram of PFR reactor and power plant.

temperature before being cleared for commissioning with steam. When steam was applied it was only a short time before some leaks were found by the sensitive detector; the leak rates being down at the 10^{-6} g/sec level. The superheaters and reheaters are of stainless steel to provide the creep properties for the steam temperature. Two of these had leaks which escalated at a reasonable rate. They were soon located by acoustic techniques. Examination showed some local cracking which subsequently was shown to be of caustic stress corrosion origin. Repairs were carried out by explosively welding on plugs, and operational techniques produced to eliminate the stress corrosion in the event of further leaks. The lower temperature conditions for CFR will not require stainless steel to be used.

Two of the evaporators, which are of ferritic steel, also had minute leaks which were much more difficult to find because they increased in size slowly and self-plugged when depressurised. Metallurgical samples were taken and it was found that the self-plugging, the form of which is indicated in Fig. 10, consists of strongly-bonded local reaction products and is impervious to helium leak detection. Much work has been done on accelerating the leak opening rate by steam and sodium conditions. The leaks have been repaired and full design operating conditions achieved.

As I have stated previously the basic sodium/water reaction work relative to fast reactor boilers has been carried out at Dounreay over many years. Initially the experiments dealt with the consequences of a large failure of steam/water tubes and the consequent reaction in the sodium. A plant design was evolved which prevents such reactions adversely affecting people and plant other than the tubes within the boilers. Experiments at full scale were carried out to prove this. Many of the features of this work are incorporated in reactors elsewhere in the world and the widely publicised events of boiler tube failure in the Russian Shevchenko reactor gave full scale operational evidence that the system is a success. It will be remembered that the incident was recorded by satellite when the hydrogen flare off from the top of the stack was seen and this became the much publicised Russian fast reactor 'accident' which is constantly being misquoted. There was no adverse affect on the plant or people and the boilers which were affected are now in operation.

In addition the whole range of leakage rates below this level has been investigated at Dounreay to determine the capability of the boiler materials to withstand local conditions from such leaks. This information is used to determine the length of continued operation in the event of a leak before shutdown of the boiler is necessary for repair.

This work has provided all the information necessary to redesign the boilers for CFR and in particular to prevent steam leaks through welds entering the sodium side. Boilers of that design using 9% Cr ferritic steel, which was successfully used for thermal reactor boilers, will be installed and tested in PFR to provide early confirmation for CFR.

The preparation for reprocessing of PFR fuel at Dounreay to extract plutonium is being carried out in close collaboration with the British Nuclear Fuel Company. Laboratory experiments on all aspects of the process have been carried out to determine the efficiency of separation. The fuel used was irradiated

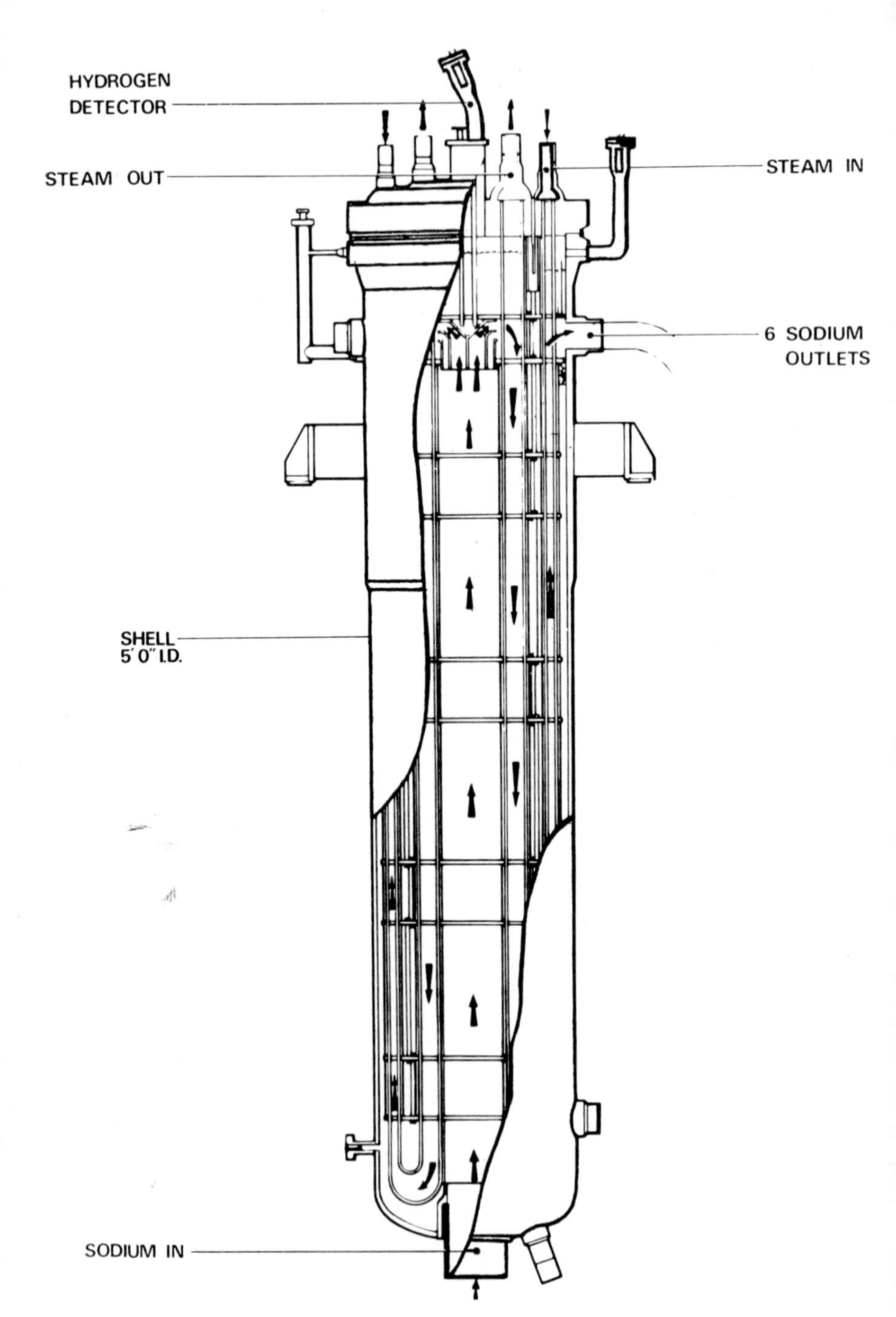

FIG. 9. Cross-section of typical PFR boiler.

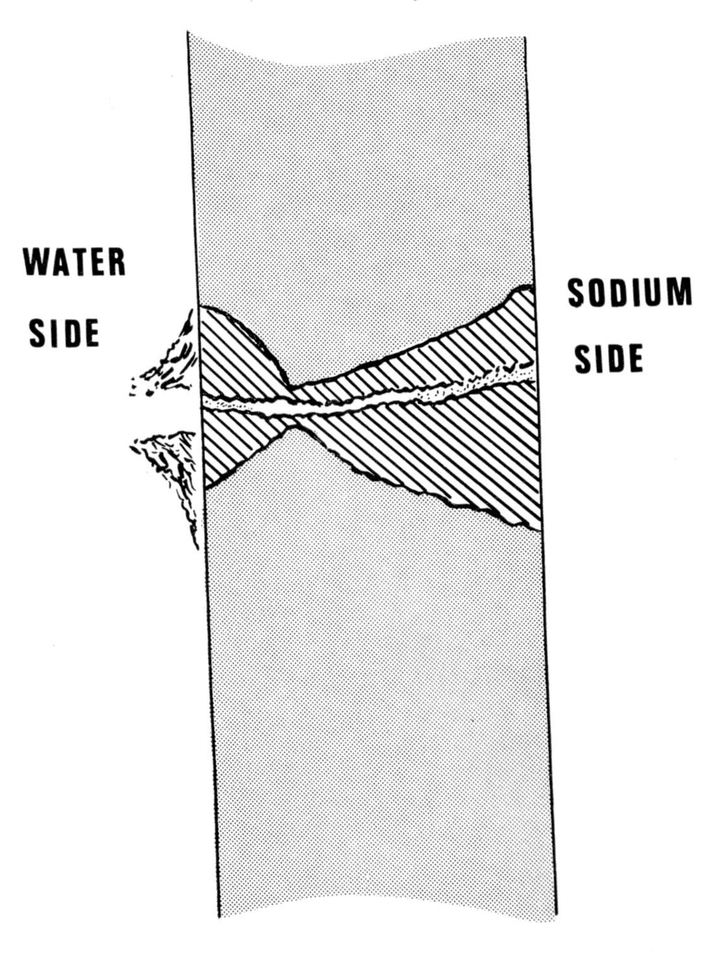

FIG. 10. Sketch of initial boiler tube leak-path self-plugged by surrounding corrosion products.

in DFR to give the correct burn-up and other physical conditions. This has ensured that all streams of the process, including waste, are understood and the safety assessments are realistic. The waste management policy has been to reduce waste to the lowest level feasible at present, to carry out experimental work to make further reductions, to hold contaminated waste in engineered retrievable stores for subsequent decontamination and recovery of any plutonium. This comprehensive work also ties in with R&D work at Windscale being carried out on incineration of combustible waste and glassification of long term waste. Consideration is also being given to the possibility of experiments in PFR to incinerate long term transuranic waste and thereby drastically shorten the half life.

Professor Hunt will talk about fuel reprocessing in the afternoon session but there is one further aspect of the fuel reprocessing work at Dounreay I should mention. It has been found that in high burn-up, high-temperature oxide fuel, the noble metals in the fission products migrate towards the central void and form various sized ingots. A typical ingot is shown in Fig. 11. During reprocessing, these and smaller particulates have to be removed from the dissolved fuel before the liquor is fed to the solvent extraction sections. The characteristics of these have been determined and taken into account in the design of the PFR fuel reprocessing plant being installed at Dounreay.

From this brief paper it will be seen that Dounreay has played an important rôle in understanding the operational aspects, safety, performance of materials and reprocessing data vital for the design of the Commercial Fast Reactor and fuel cycle.

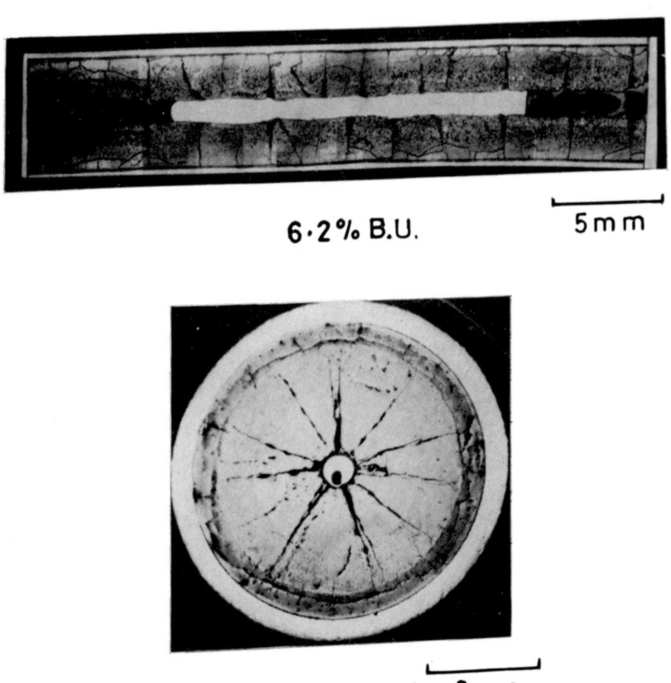

FIG. 11. Typical fission product noble metal ingots at centre void of highly irradiated oxide fuel.

The International Scene

HUGH C. SIMPSON

University of Strathclyde

SUMMARY
Six main nations or groups of nations are involved in the large scale development of Liquid Metal Fast Breeder Reactors, (1) the U.K., (2) France, (3) West Germany/Belgium/the Netherlands, (4) the U.S.S.R., (5) Japan and (6) the U.S.A. The prototype development programme of the last five of these groups is outlined. France with its 250 MW(e) Phénix plant has had two years of commercial service and has designed a 1200 MW(e) Super-Phénix plant. Germany, in collaboration with Belgium and the Netherlands, is building a 300 MW(e) plant S.N.R.-300 which is due to go critical by 1981. The U.S.S.R., in addition to the B.N.-350 combined power and desalination plant now in operation, is now building a 600 MW(e) plant B.N.-600 which is due to go critical by the end of 1977, and Japan has also started work on a 300 MW(e) reactor MONJU due to be completed by 1986. Finally, the U.S.A. in addition to a very large experimental programme, is constructing a 380 MW(e) fast reactor at Clinch River, expected to start up in 1982. This implies roughly 2300 MW of electrical power being produced by breeder reactors in the near future.

1. INTRODUCTION

In the papers presented this morning, the U.K. fast reactor scheme has been described in detail. It is my job to outline the activities of the rest of the world in breeder reactors. Inevitably, in the time available and because of the limited amount of detailed information published in the open literature, my description will tend to be superficial. The nuclear engineer will not find here plants described in sufficient detail to enable him to pass reliable judgements on their design or operation. But I hope you will get at least a strong impression of the amount of international effort going into breeder reactors and some idea of the problems involved.

Development work in breeder reactors is so costly that international and certainly inter-company collaboration is required for effective work to be done. Some of the management structures that have been developed seem almost Byzantine in complexity, and it is real credit to those involved that so much progress has been made despite these difficulties.

2. THE GENERAL PICTURE

Most of the world's industrialised nations are interested in breeder reactors, but so great is the technological and financial investment required for significant

TABLE 1

MAJOR LMFBRS IN THE WORLD

Country	*Name*	*Power (Gross Electric)*	*Operating Date*	*Primary Circuits Loop or Pool*	*Intermediate Heat Exchangers*	*Steam Generators*
U.K.	PFR	270 MW	1974	Pool/3	6	3 (1+1) (U type)
France	Phénix	250 MW	1974	Pool/3	6	3 (1+1) (S type)
Germany etc.	SNR-300	327 MW	(1981)	Loop/3	9	3 (1+1) (straight tube+helical)
U.S.S.R.	BN-350	150 MW (+desal.)	1973	Loop/6	—	—
	BN-600	600 MW	(1977)	Loop/3	6	3 (1+1) (straight tube)
Japan	Monju	300 MW	(1984)	Loop/3	3	3 (1+1) (helical)
U.S.A.	Clinch R	380 MW	(1982)	Loop/3	3	3 (2+1) (hockey stick)
		2277 MW				

progress to be made in this field that only a handful of countries or groups of countries have been capable of the effort. If we exclude the many small-scale experimental facilities for the study of reactor dynamics or sodium technology, six main national or collaborative groups can be identified, (1) the U.K., (2) France, (3) West Germany/Belgium/Netherlands, (4) the U.S.S.R., (5) Japan and (6) the U.S.A. A list of these major facilities is shown in Table 1.

Although the U.K. with the U.S.S.R. and the U.S.A. were among the first in the field in 1955, the French and the Germans are making intensive efforts to develop large scale prototype breeder reactors. Only liquid metal breeder reactors are included in Table 1. No mention is made of the gas-cooled breeder reactor, although this has its passionate advocates. However, this reactor, has not progressed much beyond the conceptual design stage.

The two U.K. breeder reactors at Dounreay have already been discussed this morning. I shall concentrate on the other reactors listed in Table 1. I am proposing to travel round the world in an easterly direction but unlike Jules Verne shall take only 25 minutes.

3. THE FRENCH BREEDER REACTORS

Despite a late start in comparison with the U.K., U.S.A. and U.S.S.R., France has had a very vigorous reactor programme since the early 1960s. Starting with a 20 MW(th) reactor built in the 1960s called Rapsodie, at Cadarache, later up-graded to 40 MW(th) in 1970, and used for the testing of fast neutron reactor fuels, the major effort in France has centred round Phénix near Marcoule on the banks of the Rhône. This is a 563 MW(th), 250 MW(e) (gross), sodium-cooled pool-type reactor not unlike the U.K. PFR described earlier. This plant went critical in August 1973 and reached full power early in 1974 (*1*). Some of the plant characteristics are shown in Table 2. It can be seen that the steam conditions at the turbine are quite conventional and that the top sodium temperature is 560°C. The core volume is also 1227 l with the fuel elements in the form of stainless steel sheathed plutonium/uranium oxide pellets. The average fuel 'rating' is 459 kW/l, very similar to the PFR.

The coolant circuits contain three primary pumps, six intermediate heat exchangers and three separate secondary loops. A diagram of the plant is shown in Fig. 1 with only one set of units shown.

After a relatively trouble-free start up (*2*), Phénix saw over two years of commercial service with a load factor of 69·4% and an availability factor of 74·4%. It experienced only temporary operating problems such as failure of the pump speed-control system, and water leaks in the steam generator intakes. Over a period of 20,000 hours, Phénix produced over 3 billion kWh (*3*). However in July and October 1976 two small secondary sodium leaks were found due to the thermal deformation of a cylindrical steel sheath in the nitrogen-filled annular space at the top of one of the intermediate heat exchangers (*4*). However, the French emphasised that this fault does not involve the reactor directly and the offending part has been removed from later plant designs. Rumour suggests that Phénix will start up again in April or May 1977, although there is, to my knowledge, no published information on this.

TABLE 2

SOME CHARACTERISTICS OF THE PHÉNIX POWER PLANT

		Phénix	*PFR*
Power	Thermal	563 MW	600 MW
	Electric (Gross)	250 MW	270 MW
	Electric (Net)	233 MW	254 MW
Fuel	Material	UO_2/PuO_2	UO_2/PuO_2
	Cladding	SS 316	SS
	Pin OD	6·6 mm	5·84 mm
	No. of Pins/Assembly	217	325
Core	Active Core Volume	1227 l	1400 l (est)
	Active Core Height	850 mm	914 mm
	Active Core Diameter	1394 mm	1400 mm (est)
Coolant	Primary Sodium In/Out	400/560°C	400/562°C
	Secondary Sodium In/Out	350/550°C	370/532°C
	Primary Sodium Flow	2760 kg/s	2923 kg/s
	Secondary Sodium Flow		
	in 3 evaporators	2210 kg/s	2923 kg/s
	in 3 superheaters	1215 kg/s	
Steam	Temperature	510°C	516°C
	Pressure	163 bar	161·7 bar
	Ratio of Thermal Power to Core Volume	459 kW/l	429 kW/l

With this very promising experience behind them, the French completed the design of a 1200 MW(e) Super-Phénix reactor to be situated at Creys-Malville in 1975. According to an agreement at the end of 1973, the power companies E. d. F. (Electricité de France) E.N.E.L. (Ente Nazionale per L'Energia Electrica) and R.W.E. (Rheinisch-Westfälisches Elektrizitätswerk) have decided to build jointly two such plants, the first in France. The design is essentially the same as Phénix, apart from the steam generator module (*3*).

It is clear that the French are well on the way to producing a large prototype plant. The operating date is reported to be 1980–81.

4. THE GERMAN/BELGIUM/DUTCH REACTOR

If we move further East into West Germany, we find that although the Federal Republic has only entered the fast reactor field in the last decade, it is now mounting a very substantial effort in collaboration with Belgium and the Netherlands. A 20 MW(e) zirconium hydride moderated reactor, KNK-2 was commissioned in Karlsruhe in 1968 for irradiation experiments, but the major collaborative effort started in 1967 on a 300 MW(e) prototype plant S.N.R. The costs are split roughly in the ratios 70: 15: 15 between Germany, Belgium and Holland.

The S.N.R.-300 is under construction at Kalkar near the German/Dutch border on the Rhine (*5*). Construction was started in 1973 but is now 18 months behind schedule (*6*), mainly because of strict licencing requirements (and the

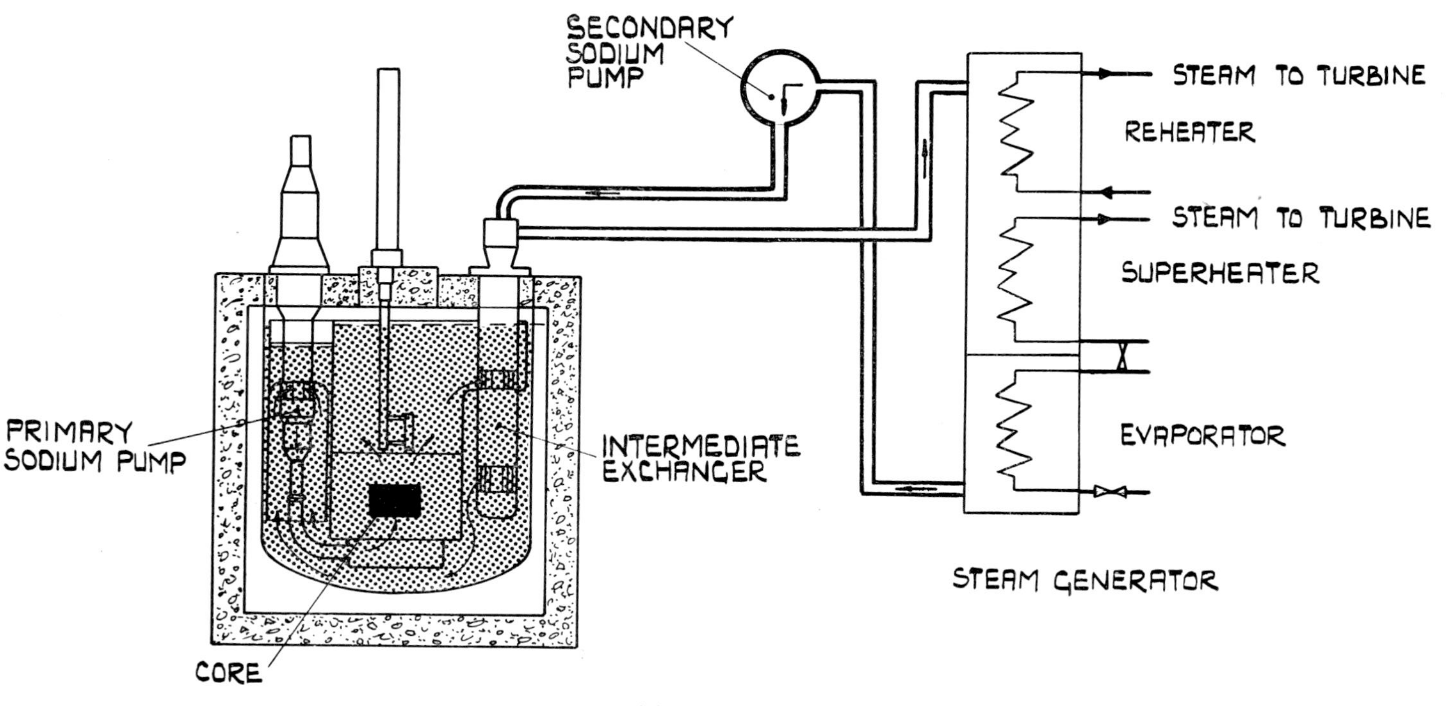

FIG. 1. Phénix reactor: pool type.

problems of international collaboration?) and is thus not likely to go critical before 1981.

The S.N.R.-300 is a sodium cooled, loop type reactor with the main characteristics shown in Table 3. The steam conditions are similar to those in Phénix, but the sodium temperatures are somewhat lower. The core volume is considerably larger although the power output is higher with 'fuel rating' 342 kW/l compared with 459 kW/l.

TABLE 3

SOME CHARACTERISTICS OF SNR-300

		SNR-300	*PFR*
Power	Thermal	762 MW	600 MW
	Electric (Gross)	327 MW	270 MW
	Electric (Net)	295 MW	254 MW
Fuel	Material	UO_2/PuO_2	UO_2/PuO_2
	Cladding	WN-1·4970 (CW)	SS
	Pin OD	6 mm	5·84 mm
	No. of Pins/Assembly	166	325
Core	Active Core Volume	2230 l	1400 l (est)
	Active Core Height	950 mm	914 mm
	Active Core Diameter	1780 mm	1400 mm (est)
Coolant	Primary Sodium In/Out	377/546°C	400/562°C
	Secondary Sodium In/Out	335/520°C	370/532°C
	Primary Sodium Flow in 3 loops	1184 kg/s/loop	2923 kg/s
	Secondary Sodium Flow in 3 loops	1088 kg/s/loop	2923 kg/s
Steam	Temperature	495°C	516°C
	Pressure	159·6 bar	161·7 bar
	Ratio of Thermal Power to Core Volume	342 kW/l	429 kW/l

The main difference between the S.N.R.-300 and Phénix is that the reactor and primary heat transfer systems are located in separate cells, instead of being integrated into a common pool as with the U.K. PFR. The primary coolant system contains 3 loops, each with its own vertical centrifugal pump, and 9 intermediate heat exchangers. The secondary coolant system also contains 3 loops with 3 evaporators. A schematic diagram of the S.N.R.-300 is shown in Fig. 2 with only one loop shown. The Kalkar plant is unique in containing a cooled floor and a core retention system, designed to take the whole of the primary sodium following a hypothetical core disruption.

The S.N.R.-300 appears to be a very conservative design with safety and licencing requirements very much in mind. The loop system was chosen to cope with the loss of coolant in case of a large pipe rupture. It is intended that the S.N.R.-300 will provide the foundation for the design and construction of larger demonstration plants. The triumvirate (E.d.F., E.N.E.L. and R.W.E.) have decided in parallel with the design of Super-Phénix to construct an equivalent plant S.N.R.-2. It is hoped to order this plant by 1982 and commission it in the late 1980s. The two European fast breeder utilities (NERSA and ESK) set up to construct Super-Phénix and S.N.R.-300 are complicated in structure, but even the C.E.G.B. has 3% interest in S.N.R.-2! (*6*).

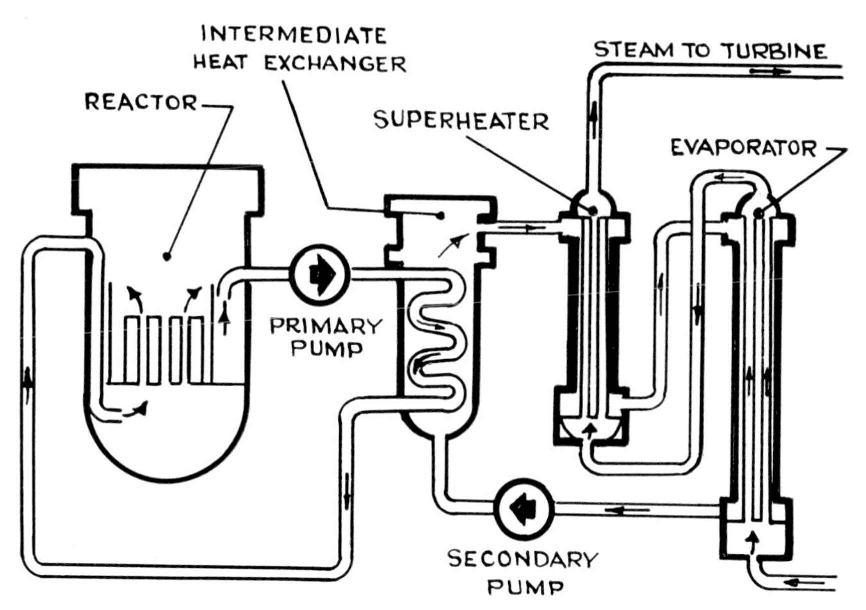

FIG. 2. S.N.R.-300 reactor: loop type.

5. REACTORS IN THE U.S.S.R.

Moving further East, we come to the U.S.S.R., one of the first countries to enter the breeder reactor field. With construction starting in the early 1950s a small 5 MW(th) plant, B.R.-5, went critical in 1958–59 at Obninsk. This early plant used plutonium oxide (PuO_2) clad in stainless steel (*7*). Following this plant was the B.O.R. 60 MW(e) experimental reactor at Melekess, construction of which started in 1963. It went critical in 1969, and had primary sodium temperatures in the quite high range of 530 to 640°C and a steam temperature of 490°C. The plant appeared to operate for several years in a relatively trouble free manner (*8*). Small sodium leakage did occur twice in the secondary circuit of the plant, once in a thermocouple sheath and once in a valve seal (*8*). On another occasion lubricating oil from the circulating pump entered the sodium and the resultant decomposition products lowered the heat transfer coefficient. However, this plant is used largely for experimental purposes e.g. in the testing of different designs of steam generators.

More ambitious is the B.N.-350 plant at Shevchanko in the Kazakh S.S.R. Construction started in 1964 and the plant was commissioned in July 1973 (*9*), representing 'the culmination of 25 years work in the U.S.S.R. in fast breeder reactors'. It is a dual purpose plant, desalting Caspian Sea water as well as producing electrical power using U 235 or plutonium enriched fuel. Of the loop type with 6 external circuits, power output and fresh water production were designed to be 150 MW(e) and 120,000 m³/day, respectively, with the reactor capacity 1000 MW(th). By 1973 the plant had reached 30% of its design capacity (*9*), with the minimum temperature of the primary sodium only 355°C. The reactor and sodium circuit was reported to operate faultlessly, although there were water leakages at welds to the sodium in the steam generator evaporators. Even as late as 1975, the B.N.-350 was constrained to 30% power operation because of problems of the sodium/steam generators (*10*). In 1974, two evaporators suffered extensive internal damage due to fast leaks; however, the safety systems operated in an entirely convincing way. The hydrogen produced was carefully flared off.

The latest Russian breeder reactor B.N. 600 is the 600 MW(e) unit being built at Beloyarsk in the Urals; it is due to go critical by the end of 1977 (*11*). It is probably the first commercial breeder reactor in the U.S.S.R., if not the world! Unlike the B.N. 350 plant, the new B.N. 600 is of pool design. It is the intention to use the experience gained in the B.N. 600 for the design of the proposed 1500 MW(e) plant.

The pool type B.N. 600 has three separate sodium circuits each containing a sodium pump and two heat exchangers. With an output of 600 MW(e) and 1500 MW(th) it has more than twice the power output of the U.K. PFR or the French Phénix with a vertical vessel size (12·8 m diameter × 13 m deep) about the same size as the British and French reactors (active core diameter is 2650 mm). The sodium pumps have twice the capacity of the British or French pumps. Thus the B.N.-600 can be regarded as a more advanced, or compact, or perhaps less conservative design than PFR or Phénix.

Another major difference is that the reactor vessel and all its internal components are supported from the bottom. In addition the Russian plant has no containment in the Western sense; instead a labyrinth of cells and rooms containing the service equipment acts as a vented 'pressure suppression containment' (*11*). Finally, a modular system of steam generator consisting of a straight tube arrangement, involving the use of tube plates top and bottom, has been chosen. In the west, tube plates in sodium/steam plants are avoided whenever possible by the use of a U-tube configuration.

It can thus be seen that the Russians have a very individualistic and active programme of breeder reactor development with designs perhaps less conservative than in the West.

6. REACTORS IN JAPAN

Moving further East again we reach Japan. The Japanese entered the nuclear reactor business much later than most other major industrialised countries. After the completion of a 100 MW(th) testing unit JOYO in 1975, work was started on a 300 MW(e) prototype power reactor MONJU (*12*).

Monju is a 714 MW(th), 300 MW(e) loop-type breeder reactor, fueled with mixed oxides of plutonium and uranium. Some details are shown in Table 4. There are three primary and secondary sodium loops leading to three sets of once-through generators producing steam at 132 bar and 487°C. With primary sodium inlet and outlet temperatures of 390°C and 540°C respectively, and top supported reactor, the design is not very different from Phénix or PFR. I understand that a considerable amount of the early preliminary engineering work was done in collaboration with the U.K.A.E.A.

TABLE 4

SOME CHARACTERISTICS OF 'MONJU'

		Monju	*PFR*
Power	Thermal	714 MW	600 MW
	Electric	300 MW	270 MW
Fuel	Material	UO_2/PuO_2	UO_2/PuO_2
	Pin OD	6·5 mm	5·84 mm
	No. of Pins/Assembly	169	325
Core	Active Core Volume	2271 l	1400 l (est)
	Active Core Height	900 mm	914 mm
	Active Core Diameter	1775 mm	1400 mm (est)
Coolant	Primary Sodium In/Out	390/540°C	400/562°C
	Secondary Sodium In/Out	320/515°C	370/532°C
	Primary Sodium Flow	3758 kg/s	2923 kg/s
Steam	Temperature	487°C	516°C
	Pressure	132 bar	161·7 bar
	Ratio of Thermal Power to Core Volume	321 kW/l	429 kW/l

The precise time scale for the construction of Monju is difficult to judge from the literature. In 1974 the Japanese reported a 1975 start to construction and a 1980 completion date (*12*). In 1975, the operating date was reported to be 1982 (*13*). In August 1976, 1984 is given as the trial operation date (*14*), but in September 1976 the construction start was reported to be 1979 with a completion date of 1986 (*15*). Whether this is because we are dealing with the incrutable East or whether this is the normal example of technological time slip, I do not know, but certainly after a late start, the Japanese are expending considerable effort in breeder reactors.

7. REACTORS IN THE U.S.A.

Finally we swing back to the Western world and into the U.S.A., the home of the first experimental breeder reactor, Clementine, operated from 1946 to 1953 at Los Alamos with only a 25 kW output. However, it was in the Experimental Breeder Reactor (EBR) series at the National Reactor Centre at Idaho Falls between 1951 and 1963 that a pioneering series of experiments were conducted leading to the detailed design of the core necessary for the safe operation of fast reactors.

The first major breeder reactor, the EBR-2, came into operation in 1964 with a 37·5 MW(th) output and this was increased to 62·5 MW(th) by 1969. This reactor was of the pot-type design, and it has continued to provide information on materials and components for over ten years, giving much of the design information on pumps, fuel handling and control built into the Clinch River prototype described below (*16*). In parallel with this, construction of the Enrico Fermi 200 MW(th) reactor was started in 1955 with funds supplied by private industry; it went critical in 1963 and produced 32×10^6 kWh before being closed down in 1973. This reactor has the distinction of suffering a melt-down of several fuel sub-assemblies, caused by a broken fairing blocking a sodium passage into the core. A somewhat touching story of the achievements of the Fermi-1 project over the 17 years is given by Alexanderson (*17*). Apart from the invaluable experience associated with coping with a partial melt down, useful information continues to be collected on the decommissioning of the breeder reactor, and the behaviour of plant components after operation for 17 years in the breeder reactor environment, experience which included sodium/water interaction in the No. 1 steam generator in 1962 (due to tube vibration) and two tube leaks in 1968 and 1971.

Another major unit in the U.S.A. under construction is the Fast Flux Test Facility (F.F.T.F.) at Hanford, Washington. It is designed as a test facility for the U.S. LMFBR programme. Although it does not produce power, it does have a capacity of 400 MW(th), with the ultimate heat rejection in sodium/air heat exchangers. Completion date was reported to be the Autumn of 1977 (*18*). It is intended to use the facility to test the fuels and materials for the LMFBRs, and, among many other objectives, to run dynamic tests up to the point of failure of the fuels, and provide operating information for the design of metals so hot that creep becomes important.

However, undoubtedly the major effort in the U.S. in LMFBRs is the construction of a demonstration plant at Clinch River near Oak Ridge. Some of the details of this plant are shown in Table 5. With a gross output of 975 MW(th) and 380 MW(e), it falls into the same size range as the S.N.R.-300. As with the German plant it is of the loop type with three separate coolant circuits. Like the S.N.R.-300, it is being designed by a triumvirate, but a private industry triumvirate this time, Westinghouse (responsible for the overall reactor and primary heat transport system), General Electric (responsible for the intermediate heat transport and steam generator systems) and Atomics International (responsible for the fuel handling, maintenance and auxiliary systems) (*19*). With excavation starting in 1974, it is expected to attain criticality in 1982 (*20*). The loop design not unlike the Westinghouse PWR plant, was chosen in preference to the pool design preferred by G.E. It will be interesting to see how G.E. works as a subcontractor under the general guidance of Westinghouse!

The Clinch River plant has primary sodium temperature in the range 388 to 534°C. The secondary sodium is fed to steam generators consisting of one superheater in series with two parallel evaporators, with steam produced at 102 bar and 482°C. The evaporators/superheaters are vertical shell and tube heat exchangers, with sodium flow vertically downward parallel to the tubes, countercurrent to the steam/water flow upwards inside the vertical tube. The evaporator/superheater is in the shape of a hockey stick to provide for differential thermal expansion between the tube and between the tube bundle and the shell.

TABLE 5

SOME CHARACTERISTICS OF THE CLINCH RIVER POWER PLANT

		Clinch R	*PFR*
Power	Thermal	975 MW	600 MW
	Electric (Gross)	380 MW	270 MW
Fuel	Material	PuO_2/UO_2	UO_2/PuO_2
	Cladding	316 SS	SS
	Pin OD	5·8 mm	5·84 mm
	No. of Pin/Assembly	217	325
Core	Active Core Volume	2400 l	1400 l (est)
	Active Core Height	914 mm	914 mm
	Active Core Diameter	1828 mm	1400 mm (est)
Coolant	Primary Sodium In/Out	388/534°C	400/562°C
	Secondary Sodium In/Out	344/502°C	370/532°C
	Primary Sodium Flow in 3 loops	1739 kg/s/loop	2923 kg/s
	Secondary Sodium Flow in 2 evaporators + 1 superheater/loop	1611 kg/s	2923 kg/s
Steam	Temperature	482°C	516°C
	Pressure	102 bar	161·7 bar
	Ratio of Thermal Power to Core Volume	406 kW/l	429 kW/l

8. CONCLUDING REMARKS

This completes our round-the-world survery of breeder reactors. The plants have been described in only a superficial way and, with the exception of the Russian plants, seem rather similar to one another apart from the choice of loop or pool design. However, the difference between a reliable plant and one liable to failure lies in the detailed engineering, particularly of the sodium heat exchangers, steam generators and sodium circulating pumps. Specialist meetings have been held dealing with the steam generators alone, with designers advocating 'hair pin' units, 'hockey stick' shaped units, double tube plate units etc. A considerable body of experience has been built up on the handling of small leaks and the maintenance of equipment, and it is probably true to say that the main outstanding design problems are in the 'plumbing' rather than in sophisticated reactor physics.

Returning again to Table 1 which summarises the international activity on LMFBRs we can see that in the not too distant future there will be roughly 2300 MW of electrical power produced by this means exclusive of Super Phénix or S.N.R.-2, which would double this figure. The breeder reactor is undoubtedly a power source of the near future and one from which Britain cannot opt out.

REFERENCES

(1) MÉGY, J., CRETTE, J. P., LABAT, P., 'Description of Phénix', *Proc. of the Int. Conf.* organised by the B.N.E.S., 'Fast Reactor Power Stations', London, March 1974, Plenary Session 1 & 2, p. 33.

(2) GUILLEMARD, B., MARECHAL, T. LE, 'Phénix Survey of Commissioning and Start-up Operations', *Proc. of the Int. Conference* organised by the B.N.E.S. etc. Plenary Session 1 & 2, p. 37. See also *Nucl. Eng. Int.*, **16,** 1971, p. 564.
(3) VENDRYÈS, G., 'International Co-operation will benefit the Development of Advanced Reactors', *Nucl. Eng. Int.*, **21,** No. 251, Dec. 1976, p. 61.
(4) MÉGY, J., 'Pourquoi Phénix est-il arrêté?' *R.G.N. Actualetés* No. 5, October–November, 1976, p. 447.
(5) MORELLE, J. M., STÖHR KARL-WALTER, VOGAL, J., 'The Kalkar Station, Design and Safety Aspects', *Nucl. Eng. Int.*, **21,** No. 246, July 1976, p. 43.
(6) BRANDSTELLER, A., EITZ, A. W., 'The Fast Breeder Programme: a Utility/Industry View', *Nucl. Eng. Int.*, **21,** No. 246, July 1976, p. 40.
(7) WOLFF, P. H. W., 'The Engineering of Fast Reactors', Bulleid Memorial Lectures, 1971, vol. V, *Engineering for Nuclear Power*, University of Nottingham.
(8) GRYAZEV, V. M., *et al.*, 'Four Years Operating Experience in the Experimental BOR 60 Nuclear Power Station', *Proc. of the Int. Conference* organised by the B.N.E.S. etc. . . . Plenary Session 1 & 2, p. 21.
(9) MITENKOV, F. M., *et al.*, 'Results of Research and Experience of Nuclear Power Station Start up with the BN-350 Reactor', *Proc. of the Int. Conference* organised by the B.N.E.S. etc. . . . Plenary Session 1 & 2, p. 27.
(10) RIPPON, S., 'Prototype Fast Breeder Reactors Operating in Europe and the U.S.S.R.', *Nucl. Eng. Int.*, **20,** No. 231, June/July 1975, p. 545.
(11) RIPPON, S., 'BN-600 Status Report', *Nucl. Eng. Int.*, **20,** No. 231, June/July 1975, p. 551.
(12) MIKI, R., SUZUKI, Y., *et al.*, 'Brief Description of Planned Prototype FBR Monju of Japan', *Proc. of the Int. Conference* organised by the B.N.E.S. etc. . . . p. 101.
(13) RIPPON, S., 'Prototype Fast Breeder Reactors Operating in Europe and the U.S.S.R.', *Nucl. Eng. Int.*, **20,** No. 231, June/July 1975, p. 550.
(14) 'AEC Emphasizes all-out efforts to Develop Fast Breeder Reactor', *The Japan Economic Journal*, Aug. 24, 1976.
(15) 'Nuclear Energy and the Electric Power Industry', *Energy in Japan*, **34,** Sept. 1976.
(16) LEMAN, J. D., BUSCHMAN, H. W., HUTTER, E., MARIARTY, K. J., 'EBR-II Ten Years Later', *Mech. Eng.*, **99,** Vol. 1, Jan. 1977, p. 33.
(17) ALEXANDERSON, E. L., 'Contributions of the Fermi-1 Project to Fast Breeder Reactor Technology', *Proc. of the Int. Conference* organised by the B.N.E.S. etc. . . . p. 13.
(18) NOLAN, J. E., MORABITO, J. J., SIMONS, A. A., COCKERMAN, D. J., 'Fast Flux Facility', *Proc. of the Int. Conference* organised by the B.N.E.S. etc. . . . p. 71.
(19) BEHNKE, W. B., CRAWFORD, J. W., JACOBI, W. M., WATSON, J. E., 'United States LFFBR Demonstration Plants Activities', *Proc. of the Int. Conference* organised by the B.N.E.S. etc. . . . p. 85.
(20) BEHNKE, W. B., Watson, J. E., NEMZEK, T. A., 'Role of the Clinch River Project in the U.S. LMFBR Program', *Nucl. Eng. Int.*, **19,** No. 221, Oct. 1974, p. 835, and other articles in this volume.

The Scottish Viewpoint

K. J. W. ALEXANDER

Chairman, Highlands and Islands Development Board

SUMMARY

At Dounreay we have one of the major successes in Highland development in the post-war world. The plant there provides direct employment for 2100 and indirect employment for probably a further 800 or more. If the decision goes against a fast reactor programme the numbers employed will fall to around 500 – an economic disaster for an area in which alternative employment is scarce and by no means easy to establish.

The development of a nuclear programme will create very significant employment in sectors of British industry which need such orders for their survival. Some important sectors of British industry will virtually disappear unless the nuclear programme moves ahead in the next few years.

Without a major expansion of nuclear-based power in Britain we must reconcile ourselves to a no growth or painfully slow growth future. If the fast reactor programme is not begun in this decade we face annual rates of growth in U.K. GDP in the early decades of next century of 2% and below. To get unemployment back to a barely 'acceptable' level of 750,000 requires an average growth rate in excess of 3% p.a.

If the U.K. decides to take the next major step in the development of nuclear power, the maximum advantage is to be gained by taking the decision sooner rather than later. Delay will lose the industrial advantages in the export field and face us with a temporary energy gap with very serious consequences. The very heavy expenditure involved in a major expansion of our nuclear programme should be spread over the period for which we enjoy the maximum benefit from North Sea oil, perhaps the only period during which the country would be able or willing to carry through such investment.

There are three reasons why the Highlands and Islands Development Board are pleased to be joint sponsors of this meeting.

Firstly, at Dounreay in Caithness we have one of the major successes in Highland development in the post-war world. The plant there provides direct employment for 2100 and indirect employment for probably a further 800 or more. If the decision goes against a fast reactor programme the numbers employed will fall to around 500 – an economic disaster for an area in which alternative employment is scarce and by no means easy to establish. On the other hand if the fast reactor programme goes ahead – employment will be stabilised at its present level; and if fast reactor were to be located at Dounreay, the Authority would at least double the numbers employed there, so that there would be around 4500 in permanent employment, with indirect employment through the local multiplier effect of a further 1500.

Secondly, the next steps in the development of a nuclear programme will create very significant employment in sectors of British industry which need such orders for their survival. Some important sectors of British industry will virtually disappear unless the nuclear programme moves ahead in the next few years. The establishment of a home-based industry supplying nuclear generating equipment, when associated with the expertise the Atomic Energy Authority and our Electricity Boards would develop, would greatly strengthen the capacity of British firms to develop a major rôle as the exporters of the equipment necessary for other countries to close their energy gaps. The list of firms which would most probably benefit includes at least five with major activities in Scotland.

Thirdly, I am convinced that without a major expansion of nuclear based power in Britain we must reconcile ourselves to a no growth or painfully slow growth future. If the fast reactor programme is not begun in this decade we must face annual rates of growth in U.K. GDP in the early decades of next century of 2% and below. Higher rates of growth can only be sustained if fast reactors make a major contribution to our energy supply before the end of this century, and given the lengthy lead times involved this requires a decision to embark on a fast reactor programme soon, certainly in this decade. These views are confirmed in the Department of Energy's own discussion document and strengthened by recent O.E.C.D. studies. The implications for employment could be particularly grave; to get unemployment back to a barely 'acceptable' level of 750,000 requires an average growth rate in excess of 3% p.a.

The most important element in any regional policy to develop the Highlands and Islands is a high rate of growth in the whole British economy. Such a high growth rate will be impossible for physical (not policy) reasons without an expanded nuclear programme, and therefore I regard such a programme as very desirable.

There is another very important contribution which an expanding nuclear segment in our energy supply can make to economic development and prosperity. The more cheaply the fuel needs of industry are met, the more efficient industry will be. Industrial expansion based on a strong competitive position which, in turn, is based on an adequate supply of energy at lower costs than would be possible by alternative methods of generation will be good for the national economy and particularly good for the regions. And, of course, there is the reasonably strong hope – increased by Lord Hinton's and Sir John Hill's remarks on Wednesday and again today – that a demonstration fast breeder reactor would be built at Dounreay. It may be of interest to report here that development projects in and around the Moray Firth which I regard as probable for the last decade of this century would consume between them approximately 300 MW of electricity additional to the present industrial demand from that area. A planned development programme for Scotland should take account of the geographical fact that plentiful fuel at relatively attractive prices could be made available close to what can become one of the major growth points not only in the Highland economy but in the Scottish economy.

However we must weigh the cost, not only in financial terms but in terms of environment, health, and security. An extra 1% on the annual rate of growth of GDP and an avoidance of very high rate of unemployment in Caithness are not to be bought at the cost of cancer, genetic mutations, and terrorism. If the scientific community were to advise that the probability of such risks occurring were, for example, one-tenth as great as the risk that if I smoke 20 cigarettes a day I shall die of lung cancer, or if I motor 2000 miles a month I shall be killed or maimed in a motor accident than I would accept that such risks were too great and the community would be best to face very slow growth, and higher unemployment.

Thus there is an additional reason for our support for this Meeting – that is to hear the views of the scientists here assembled.

Since the publication of the Sixth Report of the Royal Commission on Environmental Pollution and Lord Rothschild's article in the *The Times* last September, my impression has been that the weight of scientific opinion has not endorsed the fears expressed then and since in the public debate on whether there should now be a delay, perhaps of as long as 10 years, before a decision on a fast reactor programme is taken. It is inevitable that in a debate in which on one side are scientists whose training equips them to weigh evidence and qualify conclusions carefully and on the other side warm-hearted public-spirited citizens concerned that we should not take very terrible risks, the general public respond more to the fear of a possible disaster than to the argument that the probabilities are very heavily against the risks occurring. If we can completely avoid such occurrences by not taking the next nuclear steps, why should we take them?

But haven't we already taken the next steps? Are we not already in the plutonium age? We certainly are in the weapons field, and at Dounreay and Windscale. Interestingly enough at both places the democratically-elected public representatives have expressed themselves in favour of development. And even if Britain were to opt out of nuclear development, we would be part of a world in which nuclear developments are certain to go further. 'Stop the world I want to get off' may have been a clever title for a song but it is plaintively inadequate as an objective for decision-takers concerned for the future of a major industrial country.

Scientists do not paint the alternative scenarios in colours as vivid as those used by the opponents of nuclear development. We do not have estimates of the numbers who will be unemployed if the energy gap is not bridged, or a picture of a Britain permanently on a three-day week. How many would die of hypothermia as the cost of fuel rises relatively to the income of the retired? What social strains would arise as a result of the unsatisfied expectations of those most able to exert economic pressure for their incomes to rise much faster than the limited rate of growth of GDP?

Experience in the U.S.A. over this very hard winter has been instructive. With the total installed capacity of U.S. nuclear plants less than 20% of the system, many States depended to a much greater extent on the nuclear contribution to their power needs. For example Connecticut depended on nuclear power for 56% of its supply, Massachusetts 48%, Illinois 46%, Florida, New Jersey and Pennsylvania 39%, Vermont 28% and New York 26%. The reasons for this were the shortage of natural gas and oil and the difficulties which very low temperatures created for the handling and transportation of coal.

To summarise and conclude, if the U.K. decides to take the next major step in the development of nuclear power, the maximum advantage is to be gained by taking the decision sooner rather than later. Delay will lose us at least some of the industrial advantages in the export field. Additionally, because of the very long lead time involved in bringing into production a number of fast reactors, such delay could face us with a temporary energy gap which could have very serious consequences then and thereafter. The very heavy expenditure involved in a major expansion of our nuclear programme should be spread over the period for which we enjoy the maximum benefit from North Sea oil, benefit in government revenue and in a strong balance of payments. This may be the only period during which the country would be able or willing to carry through investment of the desired magnitude; certainly it will be the least painful period in which to bear these costs.

If the scientific community can speak with a clear, representative and therefore authoritative voice – one way or the other – it has the responsibility to do so. If it cannot, then the community may put its money on the 'something will turn up' view of the technological optimists or on the 'something will blow up' view of the environmental pessimists, and gamble with no guidance on what are the appropriate odds. I do not disguise my hope, as an economist and as an administrator concerned for the economic future of Scotland, that the scientific community will be able to speak out unequivocally in support of further nuclear advance.

Fuel Recycling

S. E. HUNT

University of Aston in Birmingham

SUMMARY

The global nuclear power programme should be designed not only to produce electricity at the lowest possible cost, but also to make the best use of our fissile fuel reserves in the longer term.

This clearly indicates the necessity of using breeder reactors, which with plutonium recycling, can achieve total fuel utilisation figures of 70% to 80% as opposed to the very small percentages available from non-breeders, even with recycling. The plutonium can be separated from spent fuel elements chemically.

The United Kingdom is in a favourable situation to initiate a fast breeder reactor programme because it has appreciable supplies of plutonium accumulated from the Magnox programme, but on a global scale there is a danger that a sudden expansion of the nuclear programme based on non-breeder reactors will exhaust the supplies of commercially viable uranium before adequate supplies of plutonium have been built up to provide the cores for a significant fast breeder programme. This situation will be worse if, as seems likely, the thermal programmes are based on reactors which are poor producers of plutonium and themselves require enriched fuel.

A more modest global expansion of the thermal nuclear programme to about 600,000 MW(e) by the year 2000 is possible using reasonably economic uranium reserves. If this were based on thermal reactors which were reasonably good producers of plutonium (Candu, Magnox and HTR) the programme could provide plutonium for the cores of 800,000 MW(e) of installed fast breeder capacity by the year 2000.

Thereafter, if the doubling time of the electricity demand is shorter than that of the plutonium inventory, either the gas-cooled fast breeder or a combination of thermal 'near breeders' and liquid-metal fast breeders will be required.

INTRODUCTION

Nuclear power appears at present to be the only large scale assured alternative to the increasing use of fossil fuels. It is, therefore, important to use the fissile fuel reserves in such a way as to provide the maximum energy in the long term, and this involves not only the use of fast breeders but the adoption of a suitable thermal reactor policy, to provide the necessary plutonium for the initial cores.

CONVERSION RATIO AND FUEL UTILISATION

A credible long term nuclear programme is dependent on the use of plutonium-fuelled breeders. This is readily appreciated if we consider the way in which the total uranium utilisation varies with conversion ratio in a system involving plutonium recycling. In the highly idealised case of 100% burn-up and plutonium recovery, the total uranium utilisation with recycling is given by: Utilisation $= X(1 + CR + CR^2 + CR^3 + \ldots CR^n)$ where X is the percentage of fissile material in the original fuel. Clearly this increases rapidly as the conversion ratio CR passes through unity.

Even making allowances for a more realistic burn-up percentage (Fig. 1) the total uranium utilisation increases dramatically from the order of a per cent or so for conversion ratios typical of present thermal reactor types to 70–80% for conversion ratios in excess of unity (*1*). It is tempting to suppose that the use of fast breeders, as opposed to low conversion ratio reactors, would increase the fuel value of our fission reserves by a factor of about 100 but this represents a gross under-estimate of their real importance.

Estimates vary, but the reserves of rich uranium ores from which uranium oxide could be recovered for a cost of say £15 per kilogram is probably limited

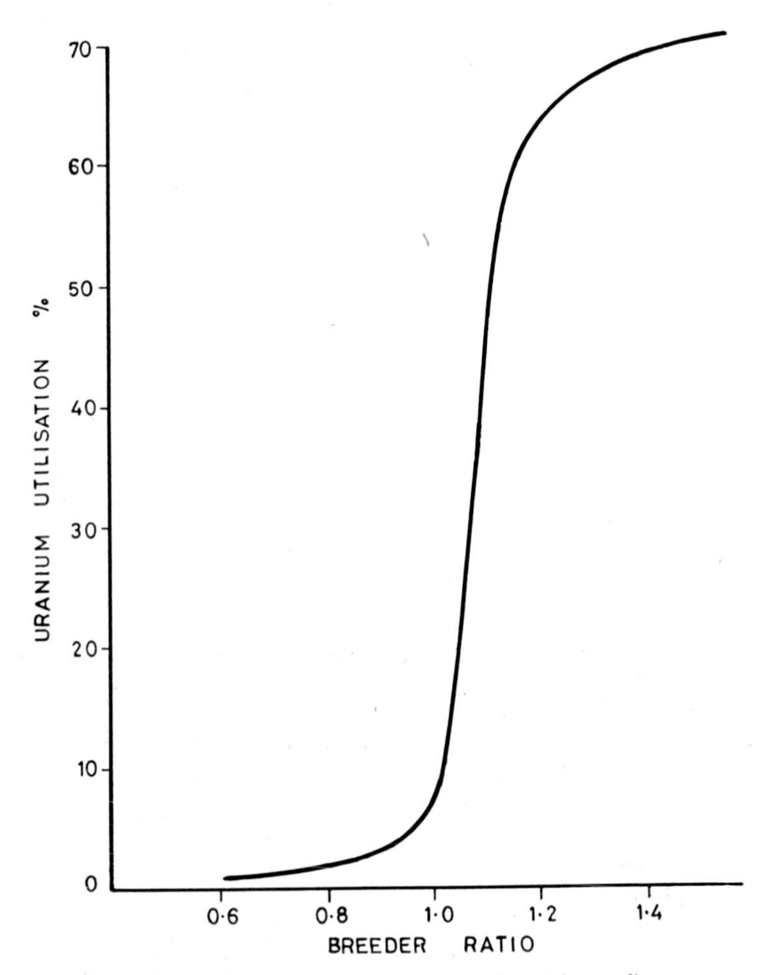

FIG. 1. Uranium utilisation vs. breeder ratio, with recycling.

to about three million tons (*2*), and if this were used in thermal reactors its fuel value would be less than 1 Q★ which is insignificant even by comparison with the known reserves of fossil fuel (*3*), (*4*). If these same reserves were used in breeder reactors their fuel value would be about 50 Q, but far more important this highly efficient use of fissile material would make many of the much poorer sources of uranium economically recoverable, and uranium in relatively dilute forms, which are consequently more costly to recover, is relatively plentiful (Fig. 2). In particular the cost of recovery from dilute sources such as

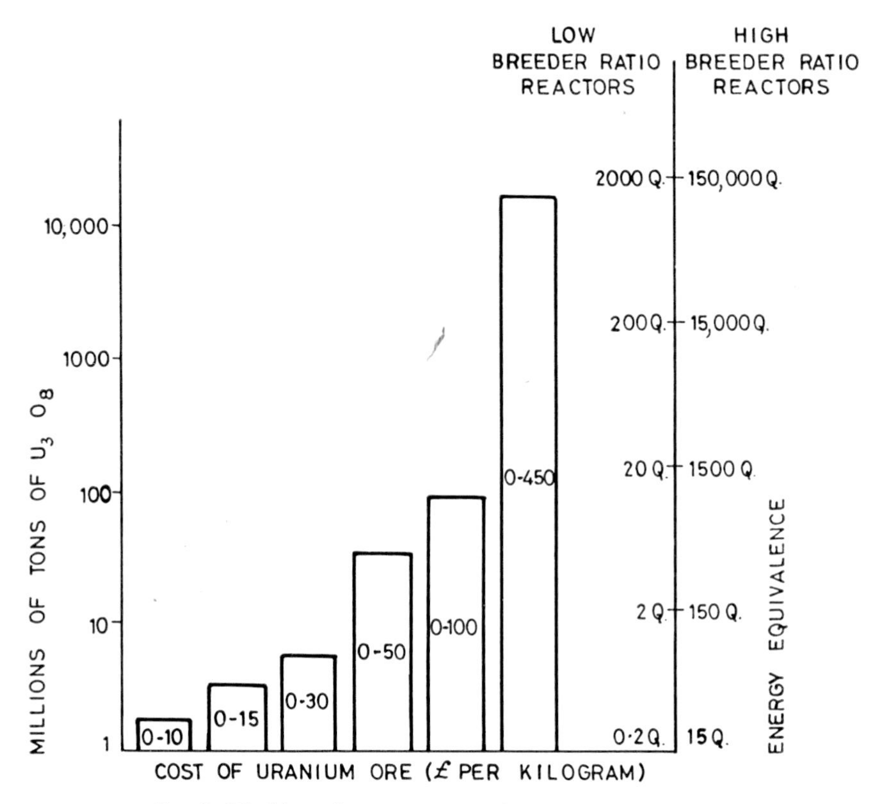

FIG. 2. World uranium reserves as a function of refining costs.

sea water has variously been estimated as between £90 and £450 per kilogram (*5*), (*6*) and these reserves, if used ultimately in fast breeders, could have a fuel value of perhaps 100,000 Q, a thousand times greater than the fossil fuel reserves. In the longer term, they could be supplemented and possibly replaced by thorium reserves using the Th-^{233}U cycle, which is entirely dependent on the use of a breeder system.

Since plutonium is the 'seed corn' for the initiation of a fast breeder programme, its recovery from spent fuel elements in the present programme is clearly important.

★ $Q = 10^{18}$ B.T.U. $\equiv 2{\cdot}9 \times 10^{14}$ kW h $\equiv 1{\cdot}05 \times 10^{9}$ T J.
This is approximately equal to the energy obtained by burning 46,500 million tons of coal.

THE RECOVERY OF PLUTONIUM

Plutonium is, of course, chemically separable from the uranium of the fuel elements in which it is produced, and from the fission products, and this is already done for the metallic Magnox fuel at Windscale with an efficiency of about 99·8%. I do not intend to describe the chemical process in detail; they involve the initial dissolving of fuel elements in nitric acid, followed by treatment by organic solvents and subsequent oxidation and purification by ion exchange processes (Fig. 3). The process has been described in detail by Walton and others (7).

It is estimated that in the U.K. we have recovered about 15 tonnes of plutonium from the Magnox programme to date. This is enough to provide the cores for almost 5000 MW(e) of commercial fast breeder capacity*, and we have, of course, accumulated vast stocks of depleted uranium for use in breeder blankets.

Once a fast breeder system is established, the actual volume of fuel for reprocessing would be very much less than that for the corresponding thermal system, but the plutonium concentration and general activity of the spent fuel would be higher. If we consider a 600 MW(e) reactor, in the fast breeder case the 5 tonne core would contain about 20% of plutonium, and the 50 tonne breeder blanket about 4%, whereas the equivalent 600 tonne Magnox core would contain between 0·1% and 0·2% of plutonium.

In the fast breeder case the separation would also be complicated by the higher concentration of ruthenium ^{106}Ru which is separated out at a relatively late stage of the process, and the care necessary to avoid the formation of supercritical volumes of fuel and liquid solvents during processing. None of these represents an insuperable difficulty, so far as I am aware, and the process does not require the large energy inputs necessary for the comparable isotope separation processes. To safeguard against the much publicised hi-jacking attempts, it seems clear that plutonium separation and re-fabrication into new fuel elements should take place on a single well-guarded site, but this does not appear to me to threaten the existence of democratic institutions.

A LONG TERM BREEDER PROGRAMME

As previously indicated the United Kingdom is well placed to initiate a modest fast breeder programme because, compared to other nations, we made an early start in our thermal reactor programme which has now been operating for the past 20 years or so, and has been based on the Magnox reactors which have the dual merit of being reasonably good plutonium producers and not themselves requiring enriched fuel.

This is not true for most other countries, however, and even if the environmental lobby were fully satisfied with the real and imagined hazards of the fast breeders, there could not at present be a significant global fast breeder

* It is assumed that 3·2 kg of plutonium are required per installed MW(e) of fast breeder capacity, half of that in the core and half in processing and re-fabrication at any one time.

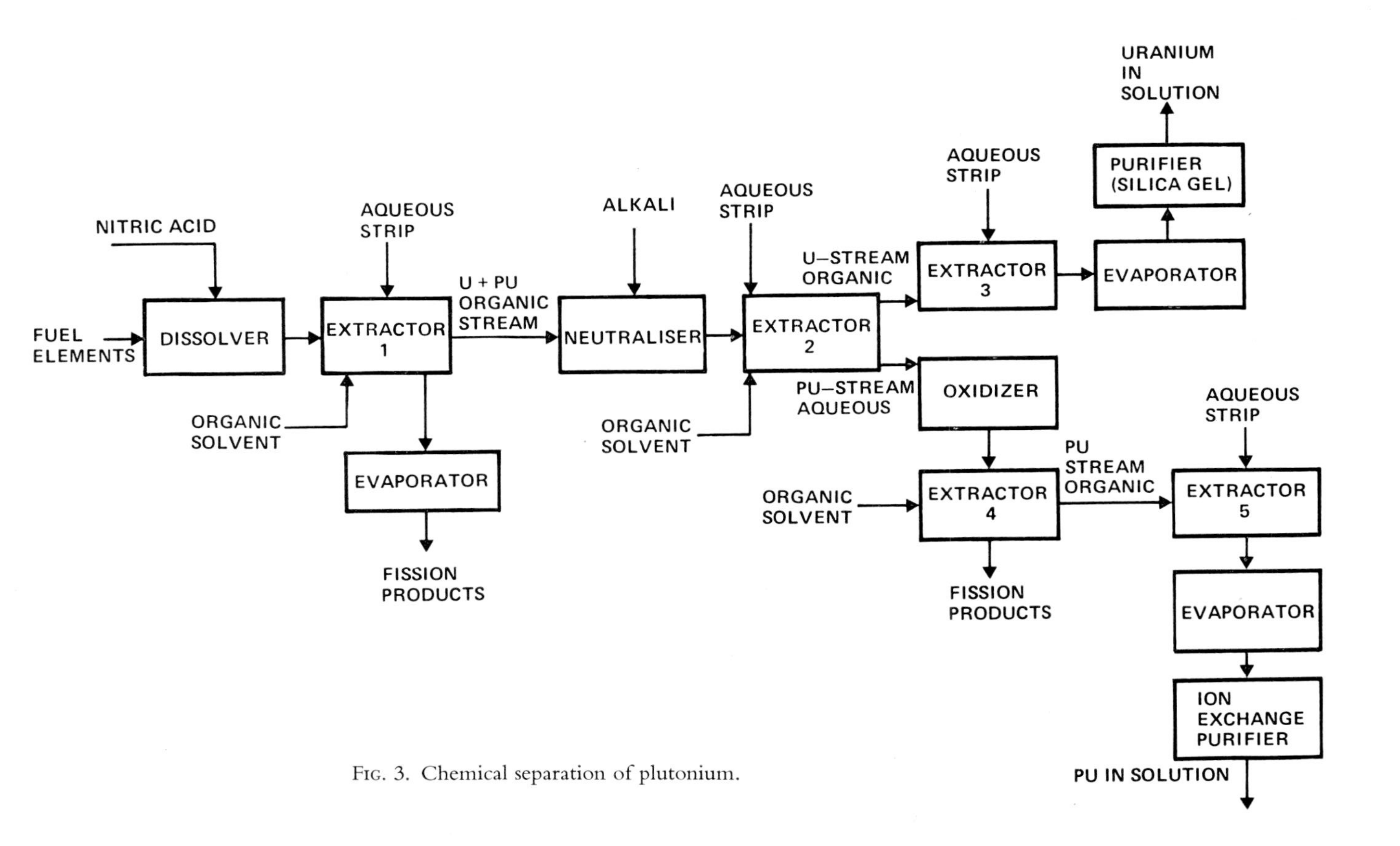

Fig. 3. Chemical separation of plutonium.

programme because adequate supplies of plutonium have not yet been produced from existing thermal programmes*, and the use of separated ^{235}U can at best be regarded as a short term palliative. Looking to the future, therefore, it seems that any thermal reactor programme must aim not only to produce electricity as economically as possible but also to accumulate an adequate supply of plutonium to mount a significant fast breeder programme before the rich reserves of uranium ore are exhausted.

This seems to indicate a gradual build-up of the thermal reactor programme based mainly on 'near breeders' which do not themselves require enriched fuel, e.g. Candu or Magnox, rather than the sudden expansion based on slightly enriched 'burner' reactors which have been dominant in the post 1970 expansion in the U.S.A., and seem likely to be prominent in the European programme (*8*).

I should like to illustrate these comments by a few very approximate calculations, dealing first with the uranium consumption by the thermal stations and then with the plutonium production from them.

URANIUM CONSUMPTION

Let us assume, for simplicity, that all thermal reactors require one tonne of natural uranium per installed MW(e) for the initial charge and that re-fuelling requires 0·3 tonnes per installed MW(e) per year. These figures are largely insensitive to the type of thermal reactor chosen (*9*). The cores of slightly enriched reactors are, of course, much smaller but the total amount of natural uranium committed via the enrichment plants is much the same.

The mean U.S.A.E.C. and E.N.E.A. projections of about 1 million MW(e) by 1990 (Fig. 4) would, on this basis, exhaust our known and estimated rich uranium reserves well before this date (Fig. 5) and a smooth extrapolation to 2 million MW(e) installed by the year 2000 would require the exploitation of poorer and more costly uranium reserves. If we assume a more modest expansion of the programme leading to 1 million MW(e) by the end of the century, our rich reserves would still be exhausted by 1995. It is only by assuming a constant installation rate of around 20,000 MW(e) per year to a total of 600,000 MW(e) by the year 2000 that our rich reserves could be extended to that date (*10*).

Here I have made the assumption that nuclear installations will be predominantly of the thermal type. It is, of course, well known that nuclear generating costs are not very sensitive to uranium prices even for thermal reactors, and that an ambitious thermal programme could probably remain economically competitive well beyond the year 2000, but it does not seem to be a good idea to exhaust our fissile reserves at this rate, unless we are assured that enough plutonium can be accumulated from the programme to enable a large scale change to a predominantly fast breeder system by then.

The various possible programmes will be examined in this light.

* Existing military stocks are assumed not to be available at present for reactor construction.

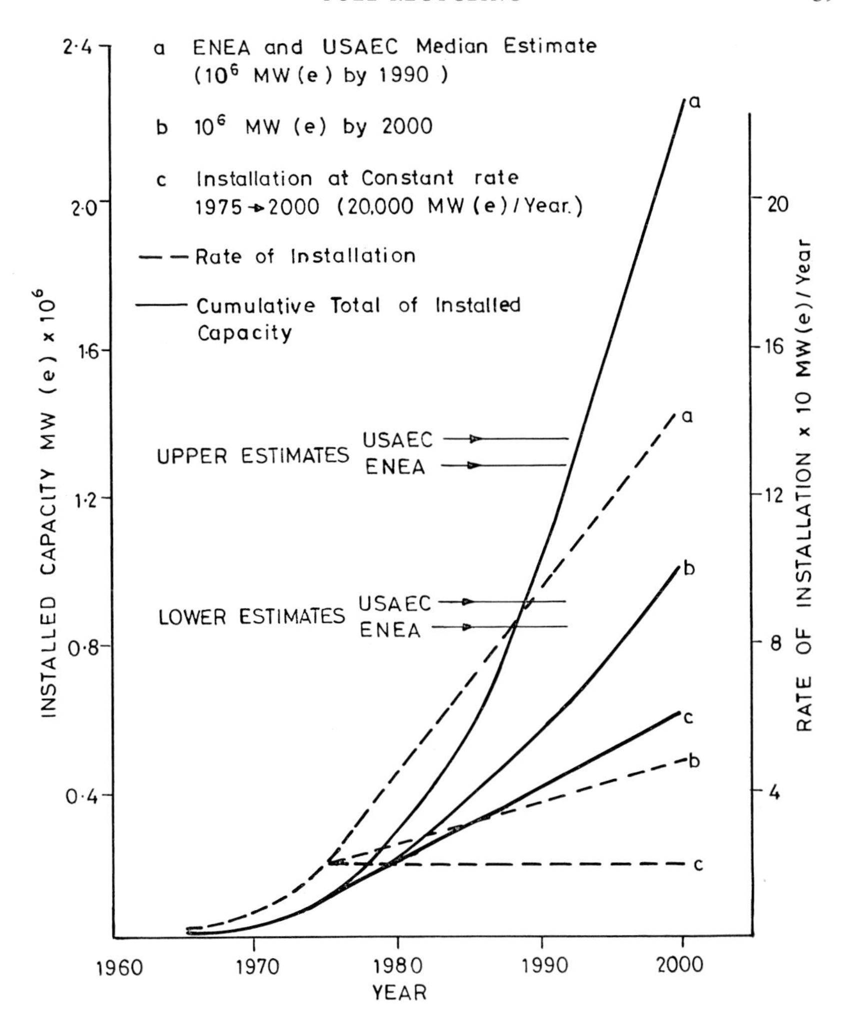

FIG. 4. Estimates of possible installed nuclear capacities by the year 2000.

PLUTONIUM PRODUCTION FROM THE THERMAL REACTOR PROGRAMMES

For simplicity, we shall divide thermal reactors into two groups, the 'near breeders', Candu, Magnox and HTR, with assumed plutonium production of 400 kg/year per 1000 MW(e) installed, and the 'burners', PWR, BWR, AGR and SGHWR with a production of 180 kg/year per 1000 MW(e) installed. The plutonium produced from each of the three postulated programmes is shown in Fig. 6, the full lines indicating the production if 'near breeders' were used and the hatched lines if 'burners' are used (as seems likely). It is seen that the rapidly expanding programme based on non-breeders(a^1) would not only exhaust the reserves of rich uranium by 1985, but would only produce about 300 tonnes of plutonium, sufficient for about 100,000 MW(e) of fast breeder capacity by then. The more gradual expansion curve based on 'near breeders' (curve c) would extend our rich reserves to 2000 and produce about 2·5 thousand tonnes of plutonium, sufficient for about 800,000 MW(e) of installed fast breeder capacity at that date. (The present world electricity generating capacity is about 1 million MW(e)). By committing the plutonium produced to fast breeders at an earlier date, say 1985 onwards, the plutonium inventory

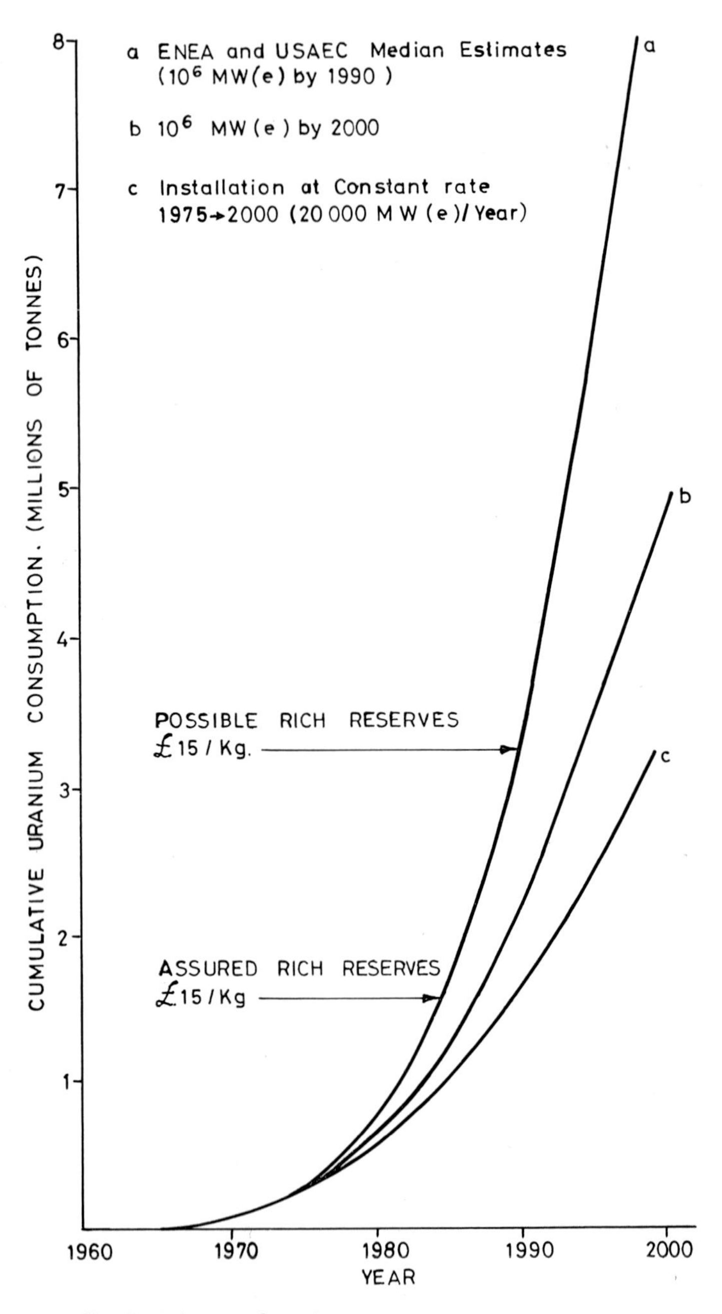

FIG. 5. Estimates of uranium consumption by the year 2000.

could be further increased as shown on the dotted curves. Conversely, if the plutonium stocks were committed to the enrichment of 'non-breeder' reactor fuel to relieve the strain on uranium separation facilities, the overall inventory would be reduced, and this appears to be a definite danger since the provision of enriched fuel is likely to be a limiting factor with slightly enriched reactor programme.

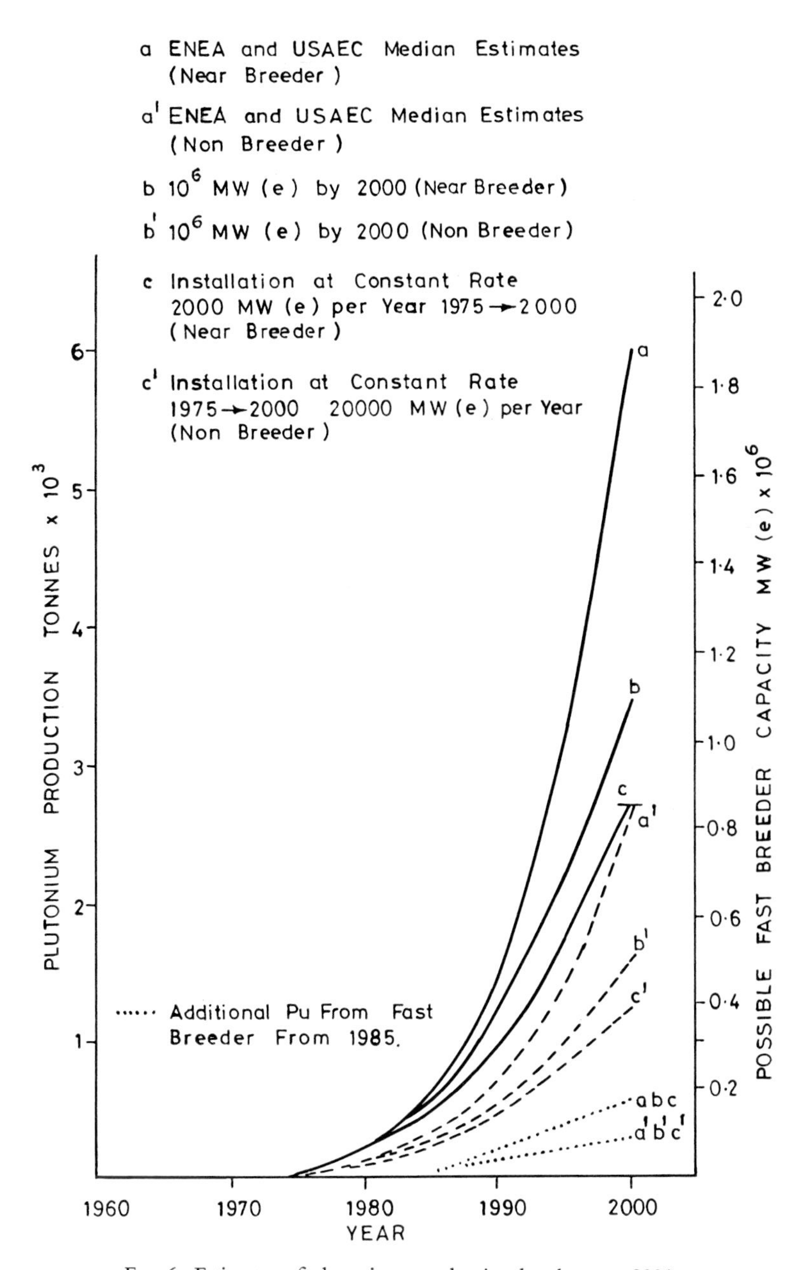

FIG. 6. Estimates of plutonium production by the year 2000.

The calculations which the curves depict are relatively crude and it has been assumed that a time delay of five years elapses between the loading of a uranium charge in a thermal reactor to the availability of the plutonium produced as fuel rods for a fast breeder. The results do appear to indicate quite clearly, however, that if sensible long term use is to be made of our fissile fuel reserves, these should be used in a relatively slow build-up of the thermal programme, preferably using 'near breeders', and the commitment of the plutonium produced to fast breeders as soon as is feasible.

Thereafter, the main question would be whether the plutonium doubling time for the fast breeders would be adequate to cope with the power demand arising from the increase in world population (and hopefully higher living standards) together with the possible shortfall in our other energy sources. It has been estimated that the present liquid-metal cooled fast breeder could approximately double the fissile material inventory every 15 to 20 years. The

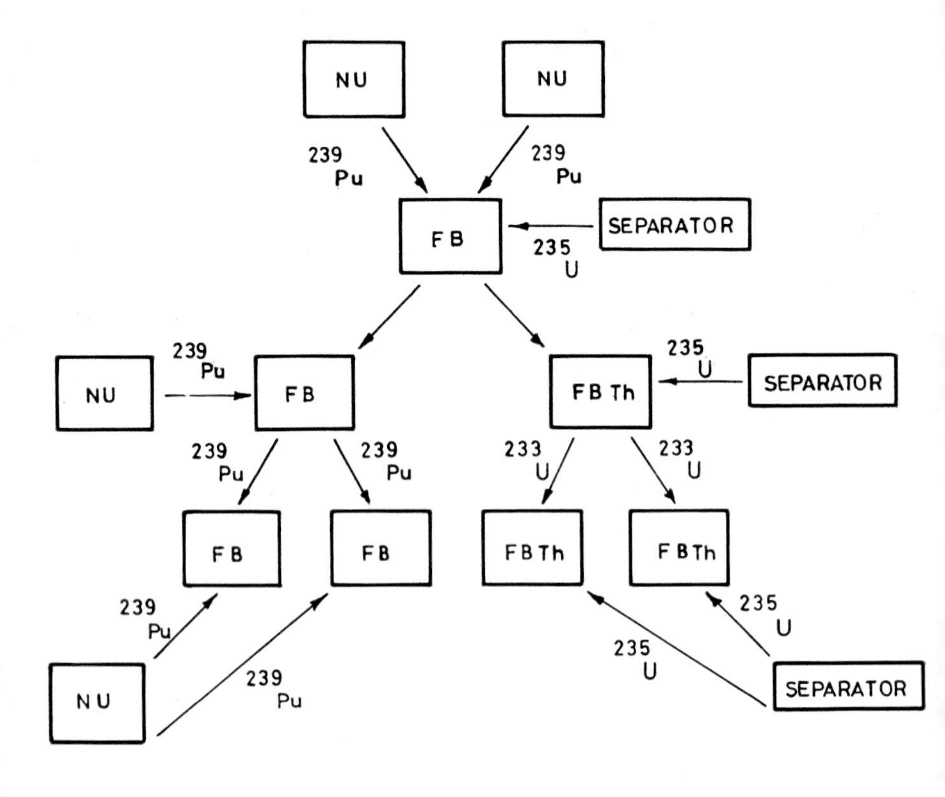

FIG. 7. A possible mixed programme for the long term.

projected gas-cooled fast breeders with their higher breeder ratio could possibly reduce the fuel doubling time to six or seven years (*11*). If this should prove to be impractical it may be necessary to supplement the plutonium supply to the fast breeders in order to keep pace with the electricity demand, and this could best be done by a 'mixed economy' of natural uranium 'near breeders' and fast breeders as shown in Fig. 7. Further supplementation of fissile material could possibly be achieved using ^{235}U from isotope separator plant, though this does not appear to be desirable from an overall energy economy viewpoint (*12*). Assuming that nuclear power is to make a significant contribution to the world long-term energy situation (and it is difficult to see credible alternatives at present) the problem is not likely to be the disposal of excess plutonium but the assuring of adequate supplies to keep the system going.

REFERENCES

(1) Seaborg, G. T., *Nuclear Energy*, Jan./Feb. (1967), p. 16.
(2) Bowie, S. H. U., *Phil. Trans. R. Soc.* A.276 (1974), p. 495.
(3) Parker, A., *Energy Policy* (1975), p. 58.
(4) Armstrong, G., *Phil. Trans. R. Soc.* A.276 (1974), p. 439.
(5) Surrey, A. J., *Energy Policy* (1973), p. 107.
(6) Marsham, T. N., Pease, R. S., 'Atom', No. 196 (1973), p. 46.
(7) *Nuclear Reactors to breed or not to breed.* (Edited by Professor J. Rotblat, Taylor & Francis, London, 1977), p. 67.
(8) Pecquer, M., *J. Brit. Nuclear Energy Society* 14, No. 1 (1975), p. 11.
(9) Surrey, A. L. *Energy Policy*, Vol. 1, 107 (1973).
(10) Hunt, S. E., *Int. J. Environmental Studies* 8 (1976), p. 235.
(11) Vaughan, R. D., *J. Brit. Nuclear Energy Society* 14, No. 2 (1975), p. 105; Slizov, V., 'The Research Programme in Nuclear Energetics at the Minsk Nuclear Power Institute U.S.S.R.', read 4th Feb. 1975, Birmingham.
(12) Hunt, S. E., *Fission Fusion and the Energy Crisis.* Pergamon Press, 1974.

The Management of Radioactive Wastes from Reprocessing Operations

D. W. CLELLAND

British Nuclear Fuels Limited

SUMMARY

The paper summarises the extensive experience gained in the U.K. over 25 years in the reprocessing of spent fuels from thermal nuclear power reactors and from dealing with the radioactive wastes. The amounts and forms of all wastes are given, with present and future management plans outlined. The risks from public exposure to the low levels of radioactivity released to the environment are examined and put into perspective.

High-level waste management is dealt with in particular detail and is demonstrated as giving rise to one centimetre cube of glass containing half gram fission products per person supplied with entirely nuclear electricity over a year, the very compact form enabling the highest standards of processing, containment and isolation to be employed.

Total releases of radioactivity to the environment from the nuclear power cycle are examined and shown controlled to very low levels, constituting only very small additions to the natural background levels. Variation due to different modes of living in different localities are demonstrated as very much more important than exposure from present or future radioactive waste disposal.

Finally, risks from nuclear power generation and radioactive waste disposal are interpreted in terms of risks to the public and shown very much less than those incurred in most common everyday activities. If the present U.K. nuclear industry is expanded ten times to supply all present electricity requirements, the risk of death from disposal of nuclear wastes is about five times less than that of being killed by lightning.

A schematic diagram of the nuclear fuel cycle is shown in Fig. 1. This shows nuclear power stations producing electricity which is sent out to consumers. Spent fuel from the reactors is sent for reprocessing and new fuel is produced in fuel element production and uranium enrichment factories.

The generation of nuclear energy, like all complex industrial activities, necessarily results in the formation of waste materials. The nature of waste generated at each stage in the nuclear fuel cycle varies considerably and each waste has to be managed in an appropriate way to ensure effective protection of the environment.

Most of the radioactivity in the nuclear fuel cycle is handled at the reprocessing plant and consequently the most significant wastes arise from reprocessing operations. Extensive experience over 25 years has been gained in

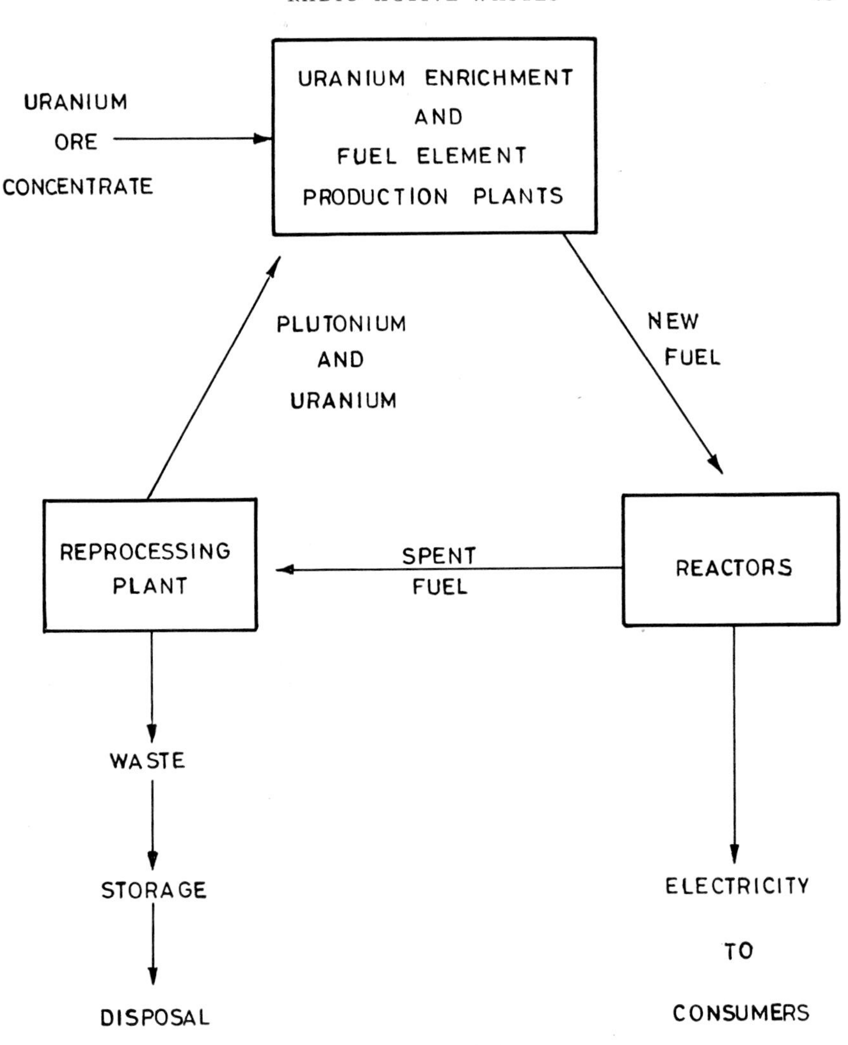

FIG. 1. The nuclear fuel cycle.

the U.K. of reprocessing spent fuel from thermal reactors and dealing with the associated range of radioactive wastes. It is this experience on which I draw for this presentation, but it is not anticipated that there will be any special problems with the wastes arising from the future reprocessing of spent fuel from fast reactors and it is envisaged that the same general management techniques will be employed and at least equivalent safety criteria will be achieved.

The average person in the U.K. at present uses about 10 kWh of electricity per day for all activities, including domestic, commercial and industrial.

On this basis, each person requires an average power output of about 400 watts and, allowing for load factor effects, an installed generating capacity of about 1–1·2 kW(e).

For every person having their present electrical needs fully supplied by nuclear power, 15 g per year of spent fuel arise from the reactors (Fig. 2).

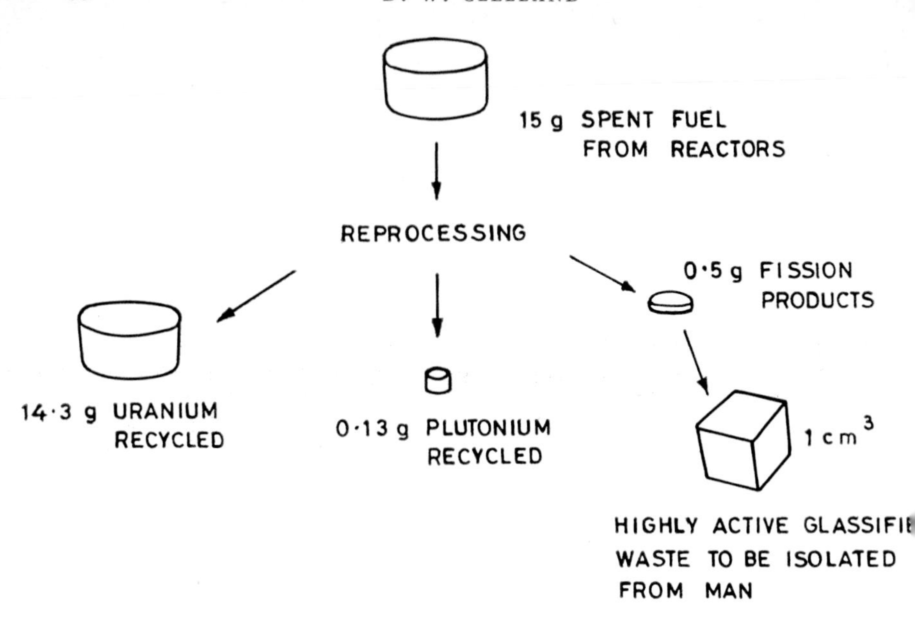

FIG. 2. Amounts of nuclear materials involved in reprocessing (per person per year).

This spent fuel comprises three main components, 14·3 g of uranium, 0·13 g of plutonium and 0·5 g of radioactive fission product wastes. The very compact form of the fuel, products and waste enables the highest standards of processing, containment and isolation to be employed.

By reprocessing this spent fuel two objectives are achieved. Firstly, the valuable materials, uranium and plutonium, are separated from the fission products and are thereby made available for the manufacture of new fuel from which more power can be generated, and secondly, the fission products which constitute a waste are conditioned for safe disposal to the environment.

Reprocessing operations consist of the following main stages:

— after reactor discharge the spent fuel is stored for about 1 year before chemical separation operations are commenced. During this period 98% of the fission product activity decays and this is a most valuable step in reducing the activity levels of wastes.
— the valuable U and Pu in the fuel are recovered for recycle.
— the wastes arising are treated according to their composition and radioactive content.

The wastes arising from reprocessing operations may be divided into two categories (Table 1). Those in the first category can be safely disposed of to the environment local to the reprocessing plant, in accordance with national and international regulations. These are disposed of routinely and do not constitute a problem.

Low-active liquid waste which arises at the rate of about 20 cm^3/d per electricity consumer, contains about one-millionth part of the activity of the spent fuel and is safely discharged to the environment local to the reprocessing

TABLE 1

RADIOACTIVE WASTES FROM REPROCESSING OPERATIONS PER ELECTRICITY CONSUMER

Wastes disposed of to local environment		*Wastes contained and isolated from man*	
— low-active liquid waste	20 cm^3/d	— highly-active liquid waste	4 cm^3/y
— low-active solid waste	50 cm^3/y	(1 cm^3/y after glassification)	
— gaseous wastes from plant ventilation	100 cm^3/min	— medium-active solid waste	25 cm^3/y
		— Pu-contaminated solid waste	0·3 cm^3/y

plant under detailed authorisations from the Ministry of the Environment and the Ministry of Agriculture, Fisheries and Food.

Low-active solid waste which arises at the rate of about 50 cm^3/y per electricity consumer, contains less than one billionth of the original activity of the spent fuel and is safely disposed of by burial in accordance with national regulations.

Gaseous waste from ventilation of the plant, which arises at the rate of 100 cm^3/min per electricity consumer, is cleansed, monitored and discharged to atmosphere in accordance with national regulations.

The second category of waste from reprocessing operations includes those wastes which contain such amounts of activity of long half-life that they must be contained and isolated from man for substantial periods of time.

Highly-active liquid waste which arises at the rate of about 4 cm^3/y for each electricity consumer contains about 1·23% of the activity of the spent fuel at discharge. This is over 99% of the activity of the fuel at the end of the 1-year storage period. At present this waste is stored in high-integrity tanks within thick-walled concrete vaults. At Windscale, some 650 m^3 of highly-active liquid waste concentrate is stored in this way. This is the cumulative total from the last 25 years of reprocessing operations.

The following management scheme is planned for this type of waste:

- storage in the liquid form for a few years
- glassification and packaging of the waste
- storage of the glassified and packaged waste in man-made structures
- final disposal of the glassified and packaged waste, after a period of cooling, into a geological formation or to the deep ocean.

Present industrial practice follows the first of these steps but, throughout the world, a very large engineering development and scientific effort has been directed to the development of glassification processes and this work is now coming to fruition.

The first industrial-scale plant for waste glassification is now completing construction and should be operational this year at Marcoule in France. It is expected that this will be followed in about five years time by another plant in France at La Hague and by a plant in the U.K. at Windscale in the mid-eighties. Other European industrial-scale glassification operations are likely to commence in the 1990s.

A considerable effort is now directed in a number of countries to the study of man-made structures which could be used for the long-term storage of highly-active waste after glassification and packaging. The following are the main alternatives under consideration:

- storage under water in ponds
- storage in air-cooled concrete vaults
- storage in underground caverns, tunnels, or holes drilled in suitable geological formations
- storage in individual packages in specially-designed flasks.

In all of these cases the waste would be in a fully-retrievable form and some surveillance would be required. Although storage under water in ponds is favoured in the U.K. the studies have already shown that any of these alternatives could be successfully applied and that the waste could be stored under such conditions for very long periods – in the timescale of centuries if desired. The immediate availability of long-term storage techniques reinforces confidence in the future safe management of these wastes since the use of such methods can ensure that sufficient time is available to explore, develop and demonstrate final disposal methods.

Although such long-term storage techniques are available which will require very little surveillance indeed, all countries with major nuclear programmes are already investigating possible methods for the ultimate disposal of the solidified and packaged waste with the object of completely eliminating any need for surveillance in the future.

Long-term storage techniques are likely to be the method employed up to about the year 2000 since disposal methods will require careful study, development and demonstration to ensure effective isolation of the wastes from man and this work will take a considerable time to complete.

The second type of waste to be isolated from man is *medium-active solid waste*. This arises from reprocessing operations at the rate of about 25 cm^3/y per electricity consumer. This waste, which contains about one hundred-thousandth part of the activity of the spent fuel at discharge, will be decontaminated, packaged and disposed of in the same way as the glassified high-level waste.

The third type for isolation, *plutonium-contaminated solid waste*, is low in beta/gamma activity but has significant alpha contamination, mainly from plutonium. It may be expected that typical operations will result in about 0·3 cm^3 per year per electricity consumer. This waste contains less than one ten-millionth part of the activity of the spent fuel and some of this waste, containing the lowest levels of alpha contamination, has been disposed of (after appropriate conditioning and packaging) to the deep ocean under internationally-organised and supervised dumping operations. For this type of waste it is intended to continue the use of sea dumping procedures under the recently ratified London Convention.

Waste containing larger quantities of plutonium is at present stored and for this a management scheme is under development which will effect plutonium recovery from the waste and condition the remainder for final disposal.

Figure 3 shows the wastes arisings from reprocessing operations with an activity balance. From this it will be seen that:

- about 98·7% of the activity of the spent fuel discharged from the reactor decays during the 1-year storage period prior to chemical separation operations. This has a major effect in reducing the amounts of activity handled and in reducing the activity of the wastes.
- about 1·23% of the activity is contained in solid wastes which are destined for long-term isolation from man. This activity represents more than 99% of the total activity in the fuel at the end of the storage period.
- the uranium and plutonium products which contain the next largest fraction of the activity 0·06% are recycled to make new reactor fuel.
- the amounts of activity released to the environment are shown in the last three items in the diagram. That these are extremely small fractions of the total activity handled indicates the care taken to minimise these quantities in accordance with national and international regulations.

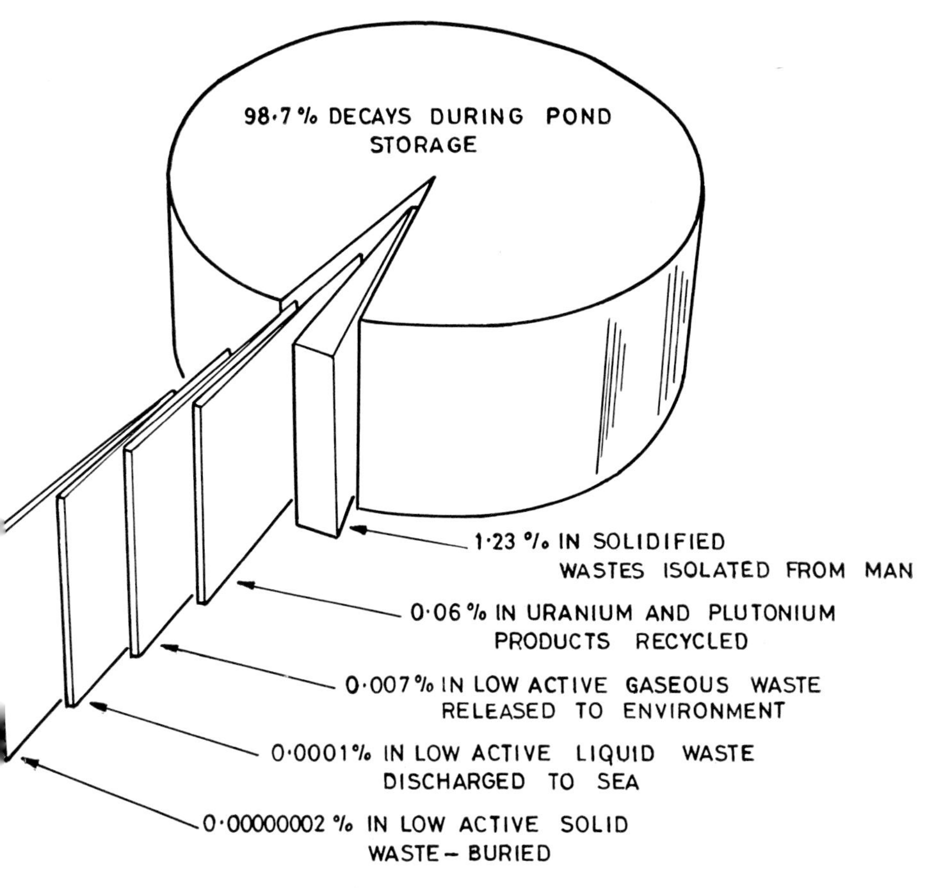

FIG. 3. Fate of activity in spent fuel reprocessing.

At this point I would like to turn to the question of the effect of waste disposal operations on the general public and I would like to try to quantify this in terms of risk and bring it into some perspective.

The environmental effect and health risks of the total nuclear fuel cycle, including waste disposal operations, have been studied by Sir Edward Pochin, a member of the U.K. National Radiological Protection Board acting as a consultant to the OECD, and the information shown in Table 2 is based on his estimates (*1*). This shows the radiation dose rate to the U.K. population from all sources and includes an estimate for the contribution of a nuclear industry which is assumed to be providing 10 kWh per day per person for the whole population. The present U.K. nuclear industry is about 1/10 of this size.

TABLE 2

RADIATION DOSE RATE TO THE U.K. POPULATION FROM ALL SOURCES
(based on estimate by E. E. Pochin, January 1976)

		Dose Rate in U.K. mrem/y
Natural background	terrestrial	30–150
	internal	24–51
	cosmic	33
Medical radiological procedures		35 (75 in U.S.A.)
Fall-out from weapons testing		6
Non-nuclear occupations		0·3
Other miscellaneous sources		0·3
Nuclear Industry providing 10 kWh/d/person	Waste disposal 0·6 Other operations 1·8	2·4
	Total, about	140

The table shows that the natural background has three components, terrestrial, internal and cosmic. The terrestrial contribution varies in the U.K. from 30–150 mrem/y depending upon whether one is living in a brick house in London or a granite house in Aberdeen. The internal radiation dose which we all receive is due mainly to the occurrence of the radioactive isotope K40 in our body tissues and varies according to the organ of the body considered.

The dose which we constantly receive from cosmic rays is fairly constant for the U.K. population at about 33 mrem/y.

The average dose in this country from radiological procedures is about 35 mrem/y.

The fall-out from weapons testing contributes about 6 mrem/y to our background, and non-nuclear occupations and other miscellaneous sources together, another 0·6 mrem/y.

By comparison it has been estimated that a nuclear industry which would provide 10 kWh/d/person for the whole U.K. population would add 2·4 mrem/y to the average total background of about 140 mrem/y. Of this 2·4 mrem/y, 0·6 mrem/y would be attributable to waste disposal. A nuclear power industry of this size and its associated waste disposal operations would therefore contribute an insignificant addition to our natural background dose.

When we compare the estimated dose levels from a nuclear industry with the variation in the components of the total natural background dose it is obvious that, for the individual, where we live and how much medical radiological examination we receive are far more significant than the contribution of a nuclear industry.

In order to obtain a further perspective on the significance of these dose levels they may be translated into units of risk and compared with other risks (*2*), (*3*) to which we are normally exposed in our daily lives.

Such a comparison is shown in Table 3. This shows the risk of death per year for an individual in the population as a result of various causes. It also includes an estimate of the risk from a nuclear industry supplying the whole population with 10 kWh/person/d.

TABLE 3

INDIVIDUAL RISK OF DEATH

Cause		*Average Risk of death per year*
All causes		1 in 83
Diseases of the circulatory system		1 in 160
All cancers		1 in 400
All accidents		1 in 3000
Motor vehicle accidents		1 in 8000
Suicide		1 in 13,000
Air travel		1 in 100,000
1½ cigarettes		1 in 1,000,000
50 miles by car		1 in 1,000,000
250 miles by air		1 in 1,000,000
1½ minutes rock climbing		1 in 1,000,000
20 minutes being a man aged 60		1 in 1,000,000
1 or 2 weeks typical factory work		1 in 1,000,000
Lightning		1 in 2,000,000
Nuclear Industry providing 10 kWh/d/person	Waste disposal	1 in 10,000,000
	Other operations	1 in 2,500,000

It will be noted that the average risk of death from all causes is 1 in 83 per year for individuals in the population and this is shown at the top of the table.

The figures below are components of this total risk which have been selected for illustrative purposes to highlight causes which will span a wide spectrum of risk. It is hoped thereby to indicate a realistic placing of the nuclear risk among other risks to which we are more accustomed.

Injury and disease may be caused by radiation as a result of the delivery of energy to sensitive body tissues and it is known that large doses of radiation given in a short time increase the risk of developing cancer, but no reliable evidence is available for low dose levels.

While the frequency of cancer induction can be estimated for absorbed doses of a few hundred rems, the frequency or even the occurrence of cancer induction at one thousandth of these dose levels is quite uncertain and can only be inferred on a hypothetical basis.

The risk shown for the nuclear industry has been calculated on the pessimistic assumption that the frequency of cancer induction is directly proportional to the radiation dose, that is, the number of rems received regardless of the size of the dose or of the rate at which it is received and is therefore a maximum estimate of the risk.

It should be noted that the risk of the nuclear industry and its waste disposal component is less to the individual than many other common activities such as, smoking 1½ cigarettes, driving 50 miles by car, or being struck by lightning.

Table 3 does not, unfortunately, show the risks of using other fuels since reliable data are not available. Ideally, when we show the risk in this table from the nuclear industry we should subtract the equivalent risk for the alternative source of energy production. If such data were available it is quite possible that an overall *reduction* in risk would be shown to be achieved by the extension of nuclear power.

I would like to complete my comments by posing the question 'Is radioactive waste disposal absolutely safe?' The purist would remind us that all activities entail some degree of risk and would reply that radioactive waste disposal constitutes a very low risk and is safer than many other common activities. Notwithstanding the accuracy of this reply I would like to submit that the average person involved in all the problems of daily living would just reply 'Yes, it is safe'.

I would now like to make a final overall summary.

So far, the stringent precautions which have been adopted in the nuclear field have been effective in protecting individuals and the environment and it is most important that, during the further expansion of our nuclear industry over the next twenty years, the present good record should be upheld.

For the future, it is most important that sufficient resources, both financial and in terms of scientific and engineering manpower, are allocated to carry into effect the waste management techniques already available, to ensure that appropriate facilities are constructed to allow efficient and safe waste management practices to be operated, and to ensure the development of acceptable disposal procedures for all categories of waste.

REFERENCES

(1) 'Estimated Population Exposure from Nuclear Power Production and other Radiation Sources' by Sir Edward E. Pochin, Nuclear Energy Agency, OECD, January 1976.

(2) *Annual Abstract of Statistics* 1976, Central Statistical Office, No. 113, 1976, HMSO.

(3) *Nuclear Power and the Environment*, Sixth Report by the Royal Commission on Environmental Pollution, para. 170, HMSO, September 1976.

Safety

J. H. FREMLIN

University of Birmingham

SUMMARY

A quantitative discussion is given of the magnitude of the risks arising from the management of the radioactive wastes from nuclear power stations. The main hazard arising from current operations in Britain is shown to be due to the liquid effluents from the processing plant at Windscale. These are run out into the Irish Sea and cause contamination of the mud of the Ravenglass Estuary and of the bottom-dwelling edible fish off the Cumbrian coast. The latter are the more important and absorb significant quantities of caesium 137. The risk to an individual eating such fish is small but if the present rate of emission of ^{137}Cs continued indefinitely the deaths from cancer of a few people per decade would eventually be produced among British consumers. This is shown to be less than one hundred-thousandth of the cancers produced in Britain by other environmental causes.

The final disposal is discussed of high-level fission-product wastes contaminated with plutonium and other actinides. The advantages of keeping these in solution under supervision in cooled tanks for a few decades is explained and the glassification for final disposal underground described.

It is pointed out that the glassified blocks will be dangerous to approach closely for a few hundred years, but that the much longer-lived alpha activity due to the actinides could affect human beings only if the glass were ground up and ingested. The blocks are insoluble in water and will therefore offer no significant risk if buried in any reasonably deep hole whether this is wet or dry.

I want to begin by discussing what we mean by safe. There is no such thing as absolute safety. I may be struck by a meteorite in the next minute. This however can be called a small risk; it is more likely that you will have to listen to the rest of my half hour; I suppose that the chance of avoiding it by this means is around 1 in 10^{18}. Even at this level, when no-one has ever observed someone in a hall in Glasgow struck by a meteorite, we can make a fair quantitative estimate of what the risk is. However, if safety meant absolute safety, there would be no point in having the word in the language. We do, and it is useful because we do not use it to mean absolute safety; we use it to mean a condition in which we believe the risks to be negligible.

It would seem sensible to neglect an avoidable risk if it were sufficiently small compared to the unavoidable risks of our lives. This is indeed an important factor. Two hundred years ago, when over most of one's life there was around $7\frac{1}{2}\%$, or 1/13 chance of death in the coming year – hence, I suppose, the superstition about not having 13 to a meal – it would have been silly to worry about a 0·1% risk of being killed in the next few years. Now, when over most of one's life one has around 0·03% chance of death in the coming year, 0·1% looks very serious indeed, but we are quite prepared to neglect risks of 1 in a million per year like walking to the shops each day without crossing a road – that is the order of risk of being killed by some vehicle mounting the pavement.

Unfortunately, there is a less sensible and more important factor, familiarity. Risks that we grow up with do not matter; risks that we did not grow up with do, independently of the size of either. Thus there is an excess of deaths from bronchial and pulmonary conditions in large European cities which must be directly or indirectly due to the fall-out from burning fossil fuels. In Britain this excess may now be decreasing but until recently has been 10 to 20,000 a year, or 100 to 200 per million tons of fossil fuel. Very few people bother, and I don't think anyone is actually frightened by this, but such a risk could not conceivably be accepted if it were new.

I hope that this example will not make anyone think I am prejudiced against fossil fuels. Without them we could not yet live in this country at all, let alone maintain the standards of life that keep most of us alive until 70 or more. I am entirely in favour of killing 20,000 people a year, usually after a fair length of life, in order that the other 55 million of us can live. But only if there is no other way.

Now at last I come to the risks of nuclear power. To justify these morally we need merely to show that a nuclear-powered economy would kill less than 10,000 of us a year. Practically, owing to its novelty, we have to do a good deal better than this. However, I hope I have made it clear that the question 'Is nuclear power safe?' is meaningless unless we ask the practical questions 'How safe?,' 'How many people will be killed?', and that the answer to this latter question cannot be zero but does have to be smaller than the number who would be killed by alternative procedures.

I have not got time to discuss all kinds of risks and I shall therefore leave the question of the risk of highly improbable but possible accidents to others and shall concentrate on the problem of radioactive wastes which are inevitably produced in normal running.

The regular emission of radioactive material from power reactors themselves is negligible; that is to say that if our whole electricity supply were derived from these the effluent would be responsible for much less than one death per century. The dangerous wastes which can reach the environment all arise from the processing of spent fuel. This might be avoided at a price if we stuck to thermal reactors but the whole point of building breeder reactors is to extract and reuse the plutonium they produce, and the processing involved inevitably leaves large quantities of radioactive waste.

Our present main processing plant in this country is that run by British Nuclear Fuels Ltd. at Windscale on the west coast of Cumbria. There the spent fuel from reactors is dissolved in nitric acid, and the uranium and plutonium separated, leaving a solution containing the main bulk of fission products. In the dissolution krypton-85, tritium and traces of ^{14}C are released and vented to the air; the yield from CFR1 running at its full rated power of 1300 MW(e) would give rise to less than one case of cancer in Britain per 1,000 years.

The main bulk of fission products is stored in double stainless steel tanks, the outer tank being monitored for leakage from the inner one and provided with pumps for returning any such leakage to a non-leaky tank. The initial activity stored is some thousands of megacuries, giving rise to perhaps 2 MW of heat, and the tanks are therefore watercooled. Multiple back-up systems make it exceedingly unlikely that the water cooling should stop; if it did do so it would take at least a few days for the tank to boil dry and in the process less than a part in 10,000 of the activity would become airborne, most of which would be deposited locally, even if the roof of the tank should be removed. This could not happen by accident and would have to happen regularly once a month for a long time, with no evacuation of local inhabitants, to kill people at the rate that they are now dying as a result of the burning of fossil fuels.

After ten years or so the short-lived isotopes producing most of the heat will have died away, but there will still be something like 100 million curies of activity per tank. At this stage it is proposed to mix the residual material with silica and other chemicals and make it into an insoluble glass. This has proved entirely satisfactory on a pilot scale but since the amount of material which can be incorporated into a ton of poorly conducting glass is limited by its heat output, the number of blocks to be made can be much reduced by waiting for at least a decade. The Windscale Harvest process would produce 15-ton blocks, one of which would be sufficient to take the annual output of CFR1.

Gamma rays from ^{137}Cs will make the blocks hazardous to approach unprotected for hundreds of years, but the actinides of which some three thousand curies will still be present in a thousand years' time are all alpha-particle emitters. The range of these particles in solids or liquids is less than 1/10 mm and the fission product hazard would have fallen to less than that of a similar volume of uranium ore. It is unfortunate that the lack of decision as to the final disposal of the blocks has led press and public to believe that no safe plan has yet been devised. In fact no decision has been made because no decision is technically needed. Any reasonably deep hole anywhere, wet or dry, would do.

It would be good for public relations to sink a shaft 1000 metres deep in rock that has been unchanged for 100 million years and later to back-fill the shaft and seal the top so that people cannot fall in. I am not claiming that this is absolutely safe. A sufficiently large and energetic band of miscreants could undoubtedly resurrect them, but fetching 15 ton blocks that you cannot approach closely up 1000 metres of rubble-filled shaft would take a little time, and when the block has to be put in the same building as anyone you wish to injure, or ground up and fed to him, it does not look a very likely proposition.

If water does percolate into the store, firstly it will leach only a few curies a year from the glass and secondly, as has been shown at Oklo, nearly all of this will be absorbed locally before the water can get anywhere near the surface. It is not always realised how much of the long-lived actinides must be put in the water supply to do any significant harm. If 100 kg of plutonium were dissolved in the Clyde, a rather odd Glaswegian who drank nothing at all but untreated Clyde water would acquire about 1/10 of the I.C.R.P. permissible body burden for members of the general public. This would indeed give some risk of cancer; equivalent to the risk presented by smoking about 1 cigarette every three weeks for one year.

The most serious effluent problem is not from the large quantity of concentrated activities but from the relatively small quantities of activity derived from washing out equipment and from the Magnox fuel storage tanks. These are delivered to the Irish Sea, and, taken up in fish, do produce a measurable hazard; until the new plant has been built this may be causing perhaps half a dozen future cancers per decade in the consumers. Much less ^{137}Cs, the main culprit, will be derived from the storage tanks for fast breeder fuel in its stainless steel or zircaloy casing. About a million people are dying per decade of cancers from environmental causes other than radiation. It would be more profitable to study these than to worry about the possible one per decade that might result from the operation of CFR1.

The Breeder Reactor in Electricity Supply

D. R. BERRIDGE

South of Scotland Electricity Board

& K. R. VERNON

North of Scotland Hydro-Electric Board

SUMMARY

Even if a decision to build a full-size fast breeder reactor were taken today, breeder reactors could not contribute significantly to Britain's energy requirements until almost the turn of the century, so today we must look ahead and examine the prospects in the year 2000.

Current annual energy usage is about 350 mtce – some 130 mt from coal itself, gas about 50 mtce, nuclear and hydro power about 15 mtce and oil about 150 mtce. By A.D. 2000 we may well require at least 500 mtce, of which our indigenous fossil fuels will meet little more than half. Coal, of which there is plenty in the ground, will be limited by its extraction rate. Indigenous oil will be a valuable export and reserves, like those of natural gas, will be waning. There will be an energy gap of between 100 and 200 mtce and what will fill it?

Tidal, solar, wind and wave power, assuming that engineering problems and amenity objections are overcome, can contribute probably no more than 40 or 50 mtce, even if economic, and dependence, therefore, on nuclear power will be substantial. Current fears about nuclear power – about waste disposal and plutonium security – will disappear as the 'unknown' becomes familiar and technology continues to advance. In world terms nuclear power will be playing a major rôle.

There will, however, have been a rapid depletion of readily available uranium ore reserves and a growing availability of plutonium from thermal reactors. Britain's resources of plutonium and depleted uranium – which the fast breeder reactor can use – will equal many thousand million tonnes of coal. Our nuclear programme should therefore include one or two FBRs.

Resources should be pooled internationally and plants built to prove alternative options and consolidate an already highly developed technology.

In Britain our earliest nuclear (Magnox) stations have served us well. In Scotland, where next year an estimated 30% of electricity output will be nuclear, Hunterston 'B' AGR has had a splendid first year of operation, and pumped storage capacity in Scotland has extended the benefits of low-cost generation.

The FBR has many very satisfactory engineering features and is eminently controllable and well behaved. It is compact, with relatively low cooling-water requirements and it could be built, one hopes, close to our load centres. There can be confidence that it will be proved safe.

An order for an FBR, on the earliest timescale that fits in with evidence of successful operation of the Dounreay PFR and an agreed international programme, would serve the national interest and ensure the survival of our plant manufacturers, so badly hit by the effects of stagnation of the U.K. economy.

If we were to decide to build a full-size demonstration commercial fast reactor today, it could hardly be designed, built and proved in service before the late 1980s. On the most optimistic prospectus it will not be until the late 1990s or the turn of the century that the FBR will be able to make a significant contribution to our energy requirements. It is a reactor, not for us (or most of us), but for succeeding generations. The question is, are they going to need it? We do not know, but we would be irresponsible if we did not try conscientiously to examine the prospects.

Sir John Hill this morning has already done this on a world energy basis. I would like to examine the position from the point of view of the Electricity Supply Industry in Britain. Although the fast reactor is essentially a producer of heat, it is through electricity that its output is most likely to be supplied to consumers.

Let us look first at overall U.K. energy requirements. Our current annual energy usage is about 350 mtce. Of this, coal itself accounts for about 130 mt, mostly and increasingly used to produce electricity. Natural gas about 50 mtce, nuclear and hydro power about 15 mtce, leaving a balance of about 150 mtce of indigenous or imported oil. We are quickly moving into a position of self-sufficiency and surplus on oil, and therefore on energy overall but, looking ahead, it seems to be generally agreed that indigenous oil reserves will be waning by the end of the century – so, too, will overseas oil resources. Meanwhile, there seems to be a very strong economic case indeed for exporting as much of our oil as we can, using as little as practicable ourselves. A lesson incidentally apparently already learned by some of the Middle East oil States who are starting to instal nuclear plants to provide home energy demand.

There is plenty of coal in the country – enough for 300 years at current rates of usage – but even with a projected massive capital investment it seems very unlikely that the rate of extraction will be very significantly increased at the end of the century – it may well be reducing.

By A.D. 2000 it seems to be generally agreed that natural gas production will be past its peak and possibly back to current production levels and still falling. It is also likely that both coal and gas will be being used increasingly as chemical feedstocks and not as primary energy sources.

Based on available figures there seems to be just a possibility that if there were no growth in energy requirements whatever by A.D. 2000 we could then still just about meet our requirements with indigenous coal, gas and oil, if reserves turn out to be towards the top end of currently predicted ranges. But there would be no oil for export and we would be on the brink of an energy shortage. We could not risk being left in that situation.

A more likely situation is that we shall need more energy – at least 500 mtce per annum by A.D. 2000 – and that indigenous fossil fuels will only be able to meet little more than half of this.

If we discount doing without or importing we are left with the prospect of needing 100 or 200 mtce per annum of something else.

Clearly there is very strong incentive to develop the so-called benign or renewable energy sources. Their prospects for technical and economic viability and the probable rate of their development are subject matter enough for a lecture in their own right. But perhaps we may attempt a very simplified assessment. Tidal power depends on suitable large estuaries or bays, of which there are two or possibly three in Britain. Together they might account for some 10 mtce per annum. Windpower would require a very great number of windmills of, say, ½ to 1 MW output each to make a significant contribution to our energy requirements. The equivalent of one large thermal power station would need several thousand mills which, with sails 70 m in diameter, would, if placed in a row, occupy several hundred miles. These and the transmission system from them would seem to present insuperable amenity problems except possibly for localised supplies.

Harnessing of wave power seems on the face of it to hold out more promising prospects but it is a long way from success in the laboratory to developing structures and machinery which can operate reliably and economically in the hostile environment of our N.W. approaches. Again the scale of any worthwhile development is vast and the engineering problems formidable. It would seem imprudent to assume that this source could contribute more than about 20 mtce by the end of the century, although it could, if all goes well, be increasing quite rapidly by that time. Solar energy is more difficult to assess since there is a range of possibilities but bearing in mind our weather and the time needed to move from a development stage to full-scale use, it seems unlikely that solar energy could contribute more than, say, another 10 mtce per annum by A.D. 2000. Thus, even if they all prove economic and practicable which is by no means certain, it seems unlikely there could be more than about 40 or 50 mtce per annum of renewable energy available to help fill the likely gap in A.D. 2000. The only realistic alternative is nuclear power, which is already developed and available. The rate of construction of nuclear power stations required during the 1990s could be very high indeed if there is much growth in overall energy requirements and if oil and gas reserves turn out to be at the lower end of the projected ranges.

Thus by the end of the century we shall already need to have a substantial dependence on nuclear power, mainly from thermal reactors. This will have demanded that public opinion is satisfied with the safety and environmental aspects of nuclear power. The debate may be long and challenging but we have no doubt that the public will be properly satisfied. Substantial neighbourhoods in this country have lived with nuclear power stations for many years. Through local liaison arrangements the people have been told about what the plants do, many people have been inside them, they know people who work in them and they are calmly accepted. Fear is almost invariably of the unknown – not the known. We all live knowingly with risks far and away higher than those from nuclear power, and it is very significant that the current fears about nuclear power are about things that are in prospect – and are therefore as yet unfamiliar to the public, i.e., new methods of waste disposal and more widespread use of plutonium. The public has of course the right to demand to be properly reassured about these aspects of nuclear power and about the issues that are raised in association with them. But by the end of the century the problems of waste disposal and plutonium security will have been seen to be contained and nuclear power, including a few fast breeder reactors, will have been accepted as a normal and necessary part of life. There is certainly nothing to suggest that because of a small additional risk the people of this country will be prepared to forgo the benefits that a plentiful supply of energy will bring them. On the contrary, woe betide the public utility or the political body that is seen to be responsible for an energy shortage. As regards the moral point that we should not leave problems in the form of nuclear waste for posterity, this argument, too, would carry little weight in the face of an energy shortage, and it is in any case as well to remember that future generations will surely have developed far more sophisticated technologies for dealing with our legacies than we can dream of. It was after all once predicted that traffic in London would eventually be brought to a halt by the rate of deposition of horse manure on its streets.

The most likely situation in Britain in A.D. 2000 will therefore be that there will already be substantial and growing dependence on nuclear power – possibly growing as much or more as a result of depletion of oil and natural gas as by growth in overall energy requirements. In world terms, too, nuclear power will be playing a major rôle. If this prognosis is correct it also follows that there will have been rapid depletion of readily available uranium ore reserves but a substantial and growing availability of plutonium from thermal reactors.

The time will be approaching, or may have been reached, when it will be possible to decide whether to go on with more nuclear power or to switch progressively to reliance on 'renewable' energy sources. The decision may well be made on purely economic grounds – but I believe our generation has a responsibility to do what it can to keep the options open – both by developing the renewable sources and by making firm provision for the continued use of nuclear power. We would be selfish indeed to burn up all the readily available uranium as well as all the oil and gas, and leave no possibility of further use of nuclear power.

The FBR provides Britain with the very attractive possibility of using what will by the end of the century have become very substantial indigenous energy resources – plutonium and depleted uranium – the equivalent of several thousand million tonnes of coal, which is almost comparable with the extent of our total coal reserves – not a prospect to be thrown away lightly.

If we accept that we are going to need to increase our use of nuclear power during the remainder of this century it must make sense to include one or two FBRs so as to ensure that a thoroughly proven and safe commercial-scale technology is available at the time when larger numbers may well be wanted rather quickly. We do not need to do this in isolation – indeed it would seem wasteful to do so. We should aim to pool knowledge and resources with other countries –

France, Germany and possibly America – and arrange between us to build a series of plants which together would prove alternative options and consolidate the technology which is already highly developed and on which, unlike most of the alternatives, the research and development has already been done.

Electricity Boards have the task of ensuring continuity of supply in both the short and the long term in a safe manner and at minimum cost. Against these criteria nuclear power has and is serving us extremely well. The early Magnox stations now up to 15 years old have proved much more reliable than any other type of generating plant. They have an outstanding safety record and their total cost of generation, including capital charges, is well below that for contemporary fossil-fired plant. The first unit of our new nuclear station at Hunterston 'B' has had an exceptionally successful first year of commercial operation and we estimate that nuclear power will generate about 30% of electricity requirements in Scotland next year. This, together with hydro power now fully developed and coal, provides the backbone of our electricity needs. In Scotland, too, we are particularly fortunate in being able to develop pumped storage in parallel with nuclear power so as to be able to get, through the day, the benefit of low-cost nuclear units generated during the period of low demand at night.

With the depletion of oil and gas reserves it seems inevitable that even greater reliance and use will be placed on electricity by the energy consumer. We must be sure we can meet this increased demand when it comes.

It seems clear to us that the further exploitation of nuclear power and pumped storage will remain the most attractive course for new generating capacity in the foreseeable future, the higher capital cost of nuclear stations being offset by the higher fuel costs of fossil-fired stations. We see the FBR as a logical and sensible development for the longer term and we have no doubt that it will serve us as well as its predecessors.

What of the FBR as an engineering proposition on our system? From what we know already from DFR, PFR and Phénix, it is clear that it has many very satisfactory features. The pool concept in which all of the main reactor components are in effect lowered in from the top and are removable for maintenance is very attractive to an operator. The fact that the reactor itself is unpressured is also a tangible advantage. We are already convinced by extensive experience that the use of liquid metal as a coolant is perfectly manageable. We have seen that despite its title of 'fast' it is eminently controllable and well behaved. We have seen, too, that even the early DFR was highly reliable – contributing its 15 MW to the Grid day after day. We know that PFR has had some initial difficulties with boiler leaks but the reasons for these are well understood and we are satisfied that as a result of the lessons learned a suitable boiler design is now available.

In almost all respects the safety argument for the FBR is simpler than for thermal reactor systems. The key issue is to ensure against the so-called fuel channel incidents referred to earlier. It seems inconceivable that such a well-defined question will not be answered with the required degree of certainty.

On siting, the FBR has good features in respect of compactness and relatively low cooling-water requirements. We hope that it will prove possible to use sites that are relatively close to our load centres rather than be driven by deterioration of our society to Nuclear Parks.

We shall of course keep our options open to exploit other means of power generation as and when they become available and we welcome the initiative that is now being shown in their development for the consumer of the future.

In the meantime failure of our national economy to grow at anything like the rates predicted in the 1960s, when new power stations now coming into service had to be ordered, means that we shall have a surplus of electricity generating plant until the mid-1980s. The ordering rate for new plant over the last 3/4 years has been almost negligible and this is causing very serious problems of survival for our generating plant manufacturers and for the nuclear industry. If we are to retain this valuable capability, and it is highly in the national interest to do so, an order for a further nuclear station must not be long delayed and reasonable continuity of orders will be needed thereafter. I have no doubt that an order for an FBR station should be included on the earliest timescale that fits in with evidence of successful operation of the PFR at Dounreay and with an agreed collaborative programme with other countries.

Summing up

F. L. TOMBS

Chairman, The Electricity Council

The early speakers pointed out the inevitability of an energy gap arising as oil and gas reserves near exhaustion, and Sir Samuel Curran drew our attention to the fact that even without growth in the world economy, demand for energy would continue to rise because development of remote sources requires increased energy investment and re-cycling of scarce raw materials also demands energy. A conservative assumption of population growth and an increase in the standard of living in under-developed countries force us to the conclusion that a substantial substitution will be needed for oil and gas. In addition to this, of course, the value of oil and gas as chemical feedstocks will increase as reserves diminish.

It is unfortunate that those who oppose the use of nuclear energy to fill the emergent energy gap usually argue on a conceptual basis and not numerically. I think that all serious observers who have attempted a numerical analysis, including potential contributions from renewable sources, are prepared to agree that a substantial nuclear contribution is necessary. It is, I believe, necessary to put objectors on the defensive by making them consider the practicability and economy of alternative energy sources.

If nuclear power is essential to make up the energy deficit, what are the grounds for the strongly voiced objections? Some concern is felt about the vulnerability of nuclear power installations to terrorists. I find myself unimpressed by this argument. If I were a terrorist and wanted to hold a population to ransom I could find an easier target with more certain and speedy adverse results on humanity and with much less personal risk to myself than would be offered by nuclear power.

Storage of long-life active waste is also a source of concern and is sometimes referred to by objectors as though we have a choice in the matter depending upon whether or not we now opt for nuclear power. In fact of course this is absurd. The problem of safe storage of long-life waste is with us from the military programme as well as from the civil nuclear programme and *has* to be solved safely. As it will be. What we are really discussing is whether or not to exploit the technical effort required to solve this problem for the benefit of man by the provision of additional energy supplies.

Proliferation of nuclear weapons is another ground for objection and featured largely in the recent Fox Report. There is really no connection, as Sir John Hill has pointed out, between the ability of determined nations to acquire nuclear weapons and the provision of nuclear power stations for civil use.

They are in no way interdependent and to deny the world benefits of civil nuclear power on these grounds would be perverse and short-sighted.

Many of the objections focus on our duty to posterity. It seems to me that posterity would be better served by conserving limited chemical feedstocks and providing for future energy needs rather than by leaving shortages in both respects to succeeding generations.

Thanks to the work done by DFR we have an impressive array of information on material, fuel performance and reprocessing. We are thus prepared in a way that no nuclear programme has previously enjoyed to build on experience and PFR will continue to add to this wealth of knowledge.

I am not myself entirely convinced that fuel reprocessing and fabrication needs to take place on the same site as power generation from breeder reactors. It seems to me that the only advantage is avoidance of transport of active materials and I would require good arguments to support the proposition that the advantages occurring from this would offset the disadvantages of cost/scale and operational flexibility.

The contributions from Professor Fremlin and Dr. Clelland provided interesting statistics on waste storage and radiation exposure of the population which served to set the problems in perspective. I found particularly forceful Professor Fremlin's arguments that nuclear power is unique in having undertaken extensive investigation of remotely possible hazards before the hazards could be shown to be active. It is as he says not surprising that the population at large should assume that nuclear power is dangerous when we put so much effort into studying risks of very low probability. Perhaps we need to demonstrate more forcefully the very substantial capital investment required to obtain reductions in small risks and to compare these with potential life-saving activities from investment in, for example, motorway crash barriers or coronary intensive-care units.

In the world at large there is widespread recognition of the need for breeder reactors and development of plant on a substantial scale in a number of countries. It seems to me a very real risk that our shortlived energy glut could be used to defer necessary action in developing nuclear power for the U.K. so that its departure may leave us in a disadvantageous position compared with countries having smaller fossil reserves who are compelled to take a more active line. We should remember, too, that we are necessarily a technology exporting country. Much of our industry is old and traditional and it would be a tragedy if we were to miss the opportunity of developing new technologies where we are not hampered by old plant and outdated attitudes. In these new technologies nuclear power surely occupies a leading position.

The need for the breeder reactor in energy terms depends on a uranium shortage and we should remind ourselves that probably only 15% of the earth's surface has been explored for uranium deposits and that not at any depth. There is a high probability that breeder reactors will be needed in numbers in the 21st century and their development therefore constitutes an essential option in our energy strategy. For this reason I believe that CFR will have to be started within the next few years in order to provide a platform from which we can expand as the energy gap approaches.

A real problem in nuclear development is that its only practical output is via the electricity supply industry which is itself in a difficult competitive position today in relation to gas. This position could be worsened by rapid introduction of inflation accounting and early adoption of long run marginal cost pricing. If these measures were to be adopted in such a way as to worsen the competitive position of electricity then the capacity of the electricity supply industry to provide the necessary capital investment in nuclear power would be greatly weakened and our ability to expand when the need arises would be impaired.

Perhaps I might close by congratulating the University of Strathclyde and the Highlands and Islands Development Board for their enterprise in organising this worthwhile meeting.

Discussion

Dr. D. J. LITTLER (*Director-General, Research Division, CEGB*)

Speaking on behalf of the CEGB I agree with the statement made by Mr. Berridge about the need for the fast reactor at the end of the century. Our studies have led us to precisely similar conclusions that there will be an energy gap and the renewable sources of energy can only partially help to fill it.

Rather than reiterate Mr. Berridge's remarks, I should like to turn to a slightly different activity which has not been mentioned so far to-day. In the CEGB we have no doubt that we cannot build a fast reactor unless it is quite safe. Indeed, we have a statutory obligation to obtain a licence for a safe system from the Nuclear Installations Inspectorate. However, we require more than this. We need a system which will have a high reliability and we are therefore undertaking studies to ensure that when we build a fast reactor, it will give a satisfactory performance. We refer to the work we do in this area as 'acting as an informed buyer' and we have teams studying the fast reactor in our Generation Division and our Nuclear Health and Safety Department, and in addition, considerable effort in the Research Division.

Our current research programme includes:

1. A range of safety studies.
   We have specialised in the interaction between fuel and coolant and have made all fast reactor protagonists pay more attention to possible dangers in this area.
2. We have a number of sodium rigs for studying:
   (a) Mass transfer, detection and control of impurities of sodium circuits and their effects on circuit activity and materials.
   (b) Instrumentation for detecting blockages, similar to the work mentioned by Mr. Blumfield.
   (c) Effects of exposure to sodium on circuit materials.
3. Fuel element behaviour including the effects of irradiation of steel by fast neutrons.
4. Structural studies – investigating, for example, the thermal shocks suffered by the fast reactor components as mentioned by Sir John Hill.

We undertake this scientific work to make us informed critics of what the Atomic Energy Authority is doing and we discuss their development work with them, if we feel that any particular technical area needs more detailed study by them. There is an iterative process and after our discussion either they change their research programme to meet our criticisms, or, of course, they convince us that such a change is not necessary. In addition to undertaking this work in our own Laboratories, we have seconded a number of staff to Dounreay to study fast reactor operation. This process works very well and I think I can say in front of the AEA people present that we have made a useful contribution to the satisfactory operation of the PFR.

I should add that at the present time there are over 60 professional staff, plus support staff, working in this area and we have had this level of effort for some years now. So that you will see that this is an additional safeguard on the satisfactory performance of the commercial fast reactor.

One aspect of our research is worthy of mention, and this is that we are undertaking some studies of the possible corrosion of candidate steels for the CFR in hot flowing sodium. We are near to commissioning rigs to do this work and the results from the rigs will give us advance information of corrosion effects, if they occur, so that we can make changes in the operation of a CFR to deal with any consequential problems.

DR. A. WARD (*The University of Strathclyde*)

Could Dr. Clelland estimate the cost of radiation waste disposal along the lines he indicates in his talk (including final burial underground) perhaps expressed as a percentage of the total cost of a unit of electrical energy?

DR. D. W. CLELLAND (*British Nuclear Fuels Ltd.*)

In reply to Dr. Ward, the cost of radioactive waste disposal will depend upon the eventual route adopted and the amount of waste accumulated before commencement of full-scale disposal, but it is expected to be of the order of a few percent of the price of electrical energy.

PROFESSOR J. H. FREMLIN (*University of Birmingham*)

With further reference to Dr. Ward's question, para. 426 of the Flowers' Report quotes an estimate by the Energy Research and Development Administration in the U.S.A., which gives 0·0033 p per kWh or less than 1% of the cost of generating electricity in Magnox reactors, now the cheapest major sources that we have. This is of course only an estimate but the main costs are for very well-established procedures and the safety level assumed to be necessary is several orders of magnitude better than that achieved by the fossil fuel industries.

MR. ROB EDWARDS (*Friends of the Earth*)

There have been so many statements made today which I would wish to question, that I am somewhat at a loss to know where to begin. Nevertheless, I will confine myself to two questions;

(a) Professor Alexander said that if we didn't go nuclear, everyone else will. I wonder if he would care to comment on President Carter's recent moratorium on domestic commercial reprocessing in the U.S. until the need for and the economics and safety of the technology are clearly demonstrated. In other words it seems at least conceivable that the U.S. could abandon nuclear energy and adopt a non-nuclear energy strategy.

(b) Another of President Carter's concerns, which has hardly been mentioned here today, is nuclear weapons proliferation. I would like to ask one of the many members of the industry here today to comment on one of the conclusions of the Fox Report (on uranium mining in Australia): 'The Nuclear Power Industry is unintentionally contributing to an increased risk of nuclear war'.

SIR JOHN HILL (*Chairman, UKAEA*)

In relation to Mr. Edwards' first point, President Carter's recent pronouncements on U.S. nuclear policy should not be interpreted as implying that the U.S. could conceivably abandon their commitment to nuclear power.

Although estimates of the rôle to be played by nuclear power in the U.S. vary, it is nevertheless recognised by President Carter that it will have an important rôle to play. The decision to call a halt to reprocessing, in that it forecloses on the option to exploit fast reactor technology with its superior utilisation of the energy potential of uranium, will therefore in the long run call for a greater number of LWRs in the U.S. to meet the required nuclear component in the U.S. energy supply mix.

The moratorium on reprocessing and the deferment of the Clinch River Breeder Reactor project should not be seen as the abandonment of fast reactor technology. As President Carter has stated, further thought is to be given to the fuel cycle to see whether a combination of reactor systems could be developed which would avoid what, in President Carter's view, are the unacceptable proliferation risks associated with the fast reactor fuel cycle as currently envisaged in the U.S. In addition ERDA is to continue to pursue a sizeable fast reactor R&D programme. We in the U.K. would not, of course, oppose the proposed reappraisal of alternative fuel cycles, since we support the non-proliferation objectives of President Carter's policy. We are, however, not convinced that the President's proposals with regard to reprocessing or to alternative fuel cycles are technically the best way of achieving those objectives.

On Mr. Edwards' second point, the nuclear power industry's unintentionally contributing to an increased risk of nuclear war, I would point out that given the world energy situation the nuclear option must be pursued if large-scale social and political upheaval is not to result due to increasing energy shortage in the future. In pursuing this option every reasonable precaution should be taken to safeguard nuclear materials against the threat of small-scale terrorist actions. Various means of minimising these risks, such as the co-location of nuclear facilities in nuclear parks, the pre-irradiation or spiking of fresh fuel elements prior to transporting them to nuclear power stations, and the stringent guarding of reprocessing facilities should be and are being looked at in detail. Taking a wider view, the possibility of governments abusing the nuclear fuel cycle for non-peaceful purposes should be guarded against by recourse to the existing safeguards system provided by the Nuclear Non-

Proliferation Treaty. In addition thought should be given to the possibility of establishing regional reprocessing centres under international control. If the services of such reprocessing centres could be assured then there would be no incentive for a large number of individual countries to establish their own reprocessing capability.

PROFESSOR J. H. FREMLIN (*University of Birmingham*)

I would reply to Mr. Edwards' two questions as follows:

(a) Whether we 'go nuclear' or not, many countries certainly will do so. Japan and several European countries, unlike the U.S.A. and ourselves, have no oil and inadequate coal supplies. None of them will risk total dependence on others for their energy supplies when the price of oil begins to increase seriously in 20–30 years' time. Neither will they take a chance on the so far unproven and rather unlikely possibility that some economic way of using solar power may be developed to a full industrial scale by that time.

The economics of processing technology are already clear for Britain. One foreign (Japanese) contract was offered covering by itself most of the capital cost of the proposed extensions. The long-term economics are established by the fact that the depleted uranium already in stock here has the energy equivalent of our entire coal reserves and the processing cost for the whole industry will be less than the cost of fuel for three or four coal-fired 1 GW power stations.

The processing plant at Windscale is responsible for less than one death per year among members of the public. This is not perfectly safe but is far safer than the fossil fuel industries whose effluent is responsible for ten thousand or so deaths, including at least some hundreds of lung cancers, every year in this country.

If the U.S.A. abandons a nuclear energy strategy it will be for political reasons or to give short term help to the oil industry; not for safety or long-term economic reasons.

(b) The generally available knowledge of the techniques of plutonium production have already made some degree of proliferation quite unpreventable. It is unlikely that the spread of this knowledge could have been stopped by any means, but whether or not this is so, the spread has not been stopped.

The rate of proliferation will however be decreased, not increased if the few countries now possessing processing plants undertake processing for those that have not, and guarantee them the fuel for their reactors. Such fuel will not then include military-grade plutonium. If Japan and other countries are denied fuel for the power stations that they regard as essential to their future existence, they must eventually set up enrichment and processing plants, treaty or no treaty. They will then have no difficulty in producing occasional batches of military-grade material.

As Dr. Eklund (Head of the International Atomic Agency, which is responsible for inspection under the Treaty) has said, President Carter's recent statement is itself inconsistent with the Non-Proliferation Treaty. If he does not withdraw or modify his proposals, therefore, it can be taken as an excuse for denunciation of the Treaty by a number of countries which are at present honouring it.

The discovery of nuclear energy has certainly enlarged enormously the damage that a war could do, though it has not necessarily made war more probable. The Fox report has given no evidence that the nuclear *power* industry has made it more likely. By providing an alternative source of energy it is in fact materially *reducing* the risk that such a war will break out in 20–30 years' time in a struggle for the remaining sources of fossil fuel.

Sir SAMUEL CURRAN (*Principal, University of Strathclyde*)

It is important to stress that all of us concerned with fission power are much distressed by the fact that plutonium provides a means of making extremely powerful weapons. For that reason among others perhaps it should be stressed that the breeder reactor is possibly the only realistic machine for consuming plutonium. Thermal reactors represent the principal means of producing plutonium in large quantities and unless an integrated system of thermal and breeder reactors is put into operation the amount of plutonium created increases steadily. It would seem unwise to leave it in spent fuel rods withdrawn from thermal reactors. It is surely more sensible to extract the plutonium and use it for peaceful purposes as fuel in breeder reactors. At the end of a few hundred years with a large integrated programme as suggested, man could eliminate all the plutonium including that in weapons. At the same time much of the uranium nearer the Earth's surface, and therefore constituting a natural hazard, would be consumed.

Professor J. S. FORREST (*University of Strathclyde*)

Before closing the meeting I would like to convey our thanks to all those who contributed to it, and especially to:

Sir Samuel Curran for initiating the meeting, providing excellent facilities to hold it and for his stimulating opening contribution;

Professor Alexander for co-sponsoring the meeting and for his interesting paper on development in the Highlands;

Lord Hinton for his words of engineering wisdom – an echo from the past and a warning for the future;

Sir John Hill and his colleagues from the Atomic Energy Authority, and all the other authors and speakers in the discussion, for their valuable contributions.

We intend to publish the proceedings of the meeting and I hope this will provide a useful record.